Alkylene Oxides and
Their Polymers

T0321095

SURFACTANT SCIENCE SERIES

CONSULTING EDITORS

MARTIN J. SCHICK
Consultant
New York, New York

FREDERICK M. FOWKES
Department of Chemistry
Lehigh University
Bethlehem, Pennsylvania

OTHER VOLUMES IN PREPARATION

Alkylene Oxides and Their Polymers

F. E. Bailey, Jr.

Union Carbide Chemicals and Plastics
   Company Inc.
South Charleston, West Virginia

Joseph V. Koleske

Consultant
Charleston, West Virginia

CRC Press
Taylor & Francis Group
Boca Raton London New York

CRC Press is an imprint of the
Taylor & Francis Group, an **informa** business

CRC Press
Taylor & Francis Group
6000 Broken Sound Parkway NW, Suite 300
Boca Raton, FL 33487-2742

First issued in paperback 2019

© 1991 by Taylor & Francis Group, LLC
CRC Press is an imprint of Taylor & Francis Group, an Informa business

No claim to original U.S. Government works

ISBN-13: 978-0-8247-8384-6 (hbk)
ISBN-13: 978-0-367-40312-6 (pbk)

**Visit the Taylor & Francis Web site at
http://www.taylorandfrancis.com**

**and the CRC Press Web site at
http://www.crcpress.com**

Preface

Alkylene oxides are interesting molecules that are used in a wide variety of ways. They are used in monomeric form as reactive intermediates to prepare low-molecular-weight chemicals that find application as solvents, pharmaceuticals, and surfactants. Ethylene oxide, for example, is used as a sterilant and as an intermediate. The utility of alkylene oxides and their polymers is far-reaching and includes the automotive, pharmaceutical, cosmetic, metal-working, mining, industrial coating, textile, construction, home furnishings, and other industries.

The poly(alkylene oxide)s are an unusual family of polymeric materials because of the extremely diverse and critical functions they perform in commercial use. To most consumers, these materials are often hidden in the fine print that describes compositions on labels, in material safety data sheets, and in handling documents. To scientists, the poly(alkylene oxide)s are known as distinct materials that are available in a wide range of molecular weights, from dimer to polymers of tens of millions, and yet remain well-characterized species of controlled molecular-weight distribution and known structure and configuration. These polymers can form ionic or molecular complexes and can provide mixtures with wide ranges of compatibilities, solubilities,

and electrical conductivities. The poly(alkylene oxide)s are in
many ways high-molecular-weight phase-transfer agents exhibit-
ing some of the host-guest specificity of the crown-ethers. The
polyether glycols have been used in this way biochemically to
permit controlled penetration of cell membranes and in frozen
cell preservation.

Commercial polymers of alkylene oxides are prepared from a
relatively few basic monomers: ethylene oxide, propylene oxide,
butylene oxide, and epichlorohydrin. (Contrary to what can be
found in the literature, poly(tetramethylene oxide) or poly-
(tetrahydrofuran) is not derived from an alkylene and is not an
alkylene oxide polymer.) The low-molecular-weight polymers are
usually liquids with viscosities that depend on composition and
degree of polymerization. These compounds are used as surfactants, lubricants, hydraulic fluids, quenchants, and emollients. The high-molecular-weight polymers are solids that may
be amorphous or partially crystalline, depending on the same
factors and on stereoregularity, with solubilities ranging from
water soluble, hydrophilic, and oil resistant to oil soluble. The
high-molecular-weight water-soluble polymers are used to modify
the flow properties of water solutions and slurries and as selective coagulants in water treatment.

Elastomers can be obtained from atactic poly(propylene oxide), polyepichlorohydrin, and epichlorohydrin/ethylene oxide
copolymers. Incorporated in polymers and plastics known as
urethanes, the poly(alkylene oxide)s are found in automobile
parts and seating, in snowmobile and other recreational vehicles,
and as motor housings for machinery. In urethane foams, they
are found in thermal insulation of appliances, in construction,
in automotive seating and padding, in home carpet underlayment,
in sports clothes, and in the upholstery of home and office
furnishings. Poly(alkylene carbonate)s prepared from alkylene
oxides and carbon dioxide are new materials that have utility as
ceramic binders and foam castings and are being investigated
for other uses.

The broad range of physical properties, chemical characteristics, and compatibilities makes these oligomers and polymers
interesting materials for researchers in the laboratory, for
product development personnel, and for consumers. They touch
our lives in some manner every day.

F. E. Bailey, Jr.
Joseph V. Koleske

Contents

v

1

Introduction: The Poly(alkylene oxide)s

I. THE POLYMERS OF THE 1,2-ALKYLENE OX-
IDES AND MAJOR COMMERCIAL USES (1—4)

The poly(alkylene oxide)s are linear or branched-chain polymers
that contain ether linkages in their main polymer chain structure
and are derived from monomers that are vicinal cyclic oxides, or
epoxides, of aliphatic olefins, principally ethylene and propylene
and, to a much lesser extent, butylene. These polyethers are
commercially produced over a range of molecular weights from a
few hundred to several million for use as functional materials
and as intermediates. Lower polymers are liquids, increasing in
viscosity with molecular weight. The high polymers can be
thermoplastic. Solubilities range from hydrophilic water-soluble
polymers that are principally derived from ethylene oxide, to
hydrophobic, oil-soluble polymers of propylene oxide and butylene oxide. A wide variety of copolymers is produced, both random copolymers and block copolymers. The latter may be used
for their surface-active characteristics.

This unusual family of materials performs extraordinarily diverse commercial functions (see Table 1). As distinct polymers,
the poly(alkylene oxide)s are of great value to polymer scien-

TABLE 1   Uses of the Poly(alkylene oxide)s
(From Refs. 3 and 4.)

Nonionic surfactants

Polyols for urethane foams and elastomers

Antifoams

Calender lubricants

Circulating oils

Compressor lubricants (air, helium, hydrogen,
    natural gas, process gases)

Cosmetics (emollients, humectants, thickeners)

Cryogenic fluids

Gear oils

Greases

Heat-transfer fluids

Hydraulic fluids

Industrial coatings

Mandrel lubricants

Metal-working lubricants

Seal lubricants

Solder-assist fluids

Textile lubricants (fiber and machine lubricants)

Two-cycle engine lubricants

Vacuum pump lubricants

---

tists who may wish to have polymers with relatively simple basic
structures that are well characterized, either linear or with
controlled branching, over a high molecular-weight range. Such
well-characterized polymers are readily prepared with narrow
molecular-weight distributions and with controlled levels of polymer crystallinity. Poly(ethylene oxide), above a molecular

weight of about 1000, is a crystalline, linear polymer with a crystalline melting point that increases with molecular weight, approaching a melting point of about 67°C. Poly(propylene oxide) and poly(1,2-butylene oxide) can be prepared as amorphous or stereospecific polymers. These polyethers can form a wide variety of molecular and ionic complexes and provide, thereby, a broad range of compatibilities, solubilities, and electrical conductivities.

Lower molecular weight polymers of ethylene oxide, and copolymers of high ethylene oxide content, are water soluble. These polymers have been used for many years in cosmetics and pharmaceuticals as humectants and binders and have been used to impart surface lubricity in lotions and creams. Poly(ethylene oxide) adducts of higher alcohols and phenols are a principal class of nonionic surfactants. A higher polymer, CARBOWAX^{TM} 20M, with a number average molecular weight of about 20,000, has been widely used as a packing in gas chromatograph columns.

High-molecular-weight polymers of ethylene oxide are miscible with water in all proportions at temperatures from the freezing point to near the boiling point, depending on polymer molecular weight. The polymers exhibit an inverse solubility-temperature relationship in water, having a lower consolute temperature near 100°C. At very high dilutions, poly(ethylene oxide) is an enormously effective drag-reduction agent in water (5) (Tom's Effect). At higher concentrations, it forms hydrogels. The polymer itself can be formed as a thermoplastic into a water-soluble film.

The lower molecular weight polymers of propylene oxide and propylene oxide adduct copolymers are used as surfactants, hydraulic fluids, and machine and metal-working lubricants. Propylene oxide adducts of polyhydroxy compounds, such as glycerine, trimethylolpropane, or pentaerythritol or sorbitol, are principal polyols used in making polyurethane elastomers, rigid thermal insulation, flexible foams, and coatings.

The poly(ethylene oxide)s and poly(propylene oxide)s described in this book are widely available in different molecular weights and functionalities. Table 2 is a listing of some of the various trade names or brand names and their sources for these polymers and their copolymers. Trade names for other products can be found at various appropriate places within the text.

TABLE 2   Trade Names or Brand Names for Poly(ethylene oxide)s and Poly(alkylene oxide)s and Their Copolymers

Trade name	Source
ACTOL	Allied-Signal Incorporated
ADEKA, ADEKA CARPOL	Asahi Denka Kogyo K.K.
AFTOL	Petrolite Corporation
ALKATRONIC	Alkaril Chemicals Inc.
ANTAROX	Sparmar Dispersants
AQUANOX	Baker Hughes Incorporated
AQUASURF	Baker Hughes Incorporated
ATLAS	Atlas Chemical Company
CP POLYOL	Corn Products Incorporated
CARBOWAX	Union Carbide Corporation
CARPOL	E. R. Carpenter, Incorporated
CARDANOL	Shell Chemical Company
DALTOLAC	ICI Limited
DESMOPHEN	Farbenfabriken Bayer A.G.
GAFTRONIC	GAF Corporation
EMERLUBE	Quantum Chemical Corporation
FOAMBREAK	Baker Hughes Incorporated
FOMREZ	Witco Corporation
HODAG	Hodag Chemical Corporation
ISONOL	Upjohn Company
JEFFOX	Texaco, Incorporated
MULTRANOL	Bayer U.S.A., Mobay Corp.
NALCO	Nalco Chemical Compay
NIAX	Union Carbide Corporation
PLURACOL	BASF Corporation

TABLE 2 (Continued)

Trade name	Source
PLURONIC	BASF Corporation
POLY-G	Olin Corporation
PROPYLAN	Lankro Chemical Company, Mitsui K.K.
POLYOX	Union Carbide Corporation
POLYURAX	Pfizer Incorporated
SELECTRO	PPG Ind., Incorporated
SYNPERONIC	ICI American Holdings, Inc.
TETRONIC	BASF Corporation
THANOL	Arco Chemical Company
TRETOLITE	Petrolite Corporation
VISCO	Nalco Chemical Company
VORANOL	Dow Chemical U.S.A.
WITBRAKE	Witco Corporation

REFERENCES

1. F. E. Bailey, Jr., The Chemist 9:9 (July-August 1987).
2. F. E. Bailey, Jr., in Encyclopedia of Materials Science and Eng. (M. B. Bever, ed.), Pergamon Press, Oxford, 1986, p. 3647.
3. S. A. Cogswell, in Chemical Economics Handbook, SRI International, Menlo Park, 1986.
4. A. O. Masilungan, in Chemical Economics Handbook, SRI International, Menlo Park, 1984.
5. P. N. Georgelos and T. M. Torhelson, J. Non-Newt. Fluid Mech. 27:191 (1988).

2
Alkylene Oxides: Manufacture, Chemistry, and Applications

I. ETHYLENE OXIDE

World annual production of ethylene oxide is about 13 billion pounds (5.5 million metric tons). About 44 percent of this production is in the United States, about 25 percent is in Western Europe, and 9 percent is in Japan. The principal uses for ethylene oxide in the United States are listed in Table 1.

The principal uses of ethylene oxide (about 60 percent) are as automotive antifreeze and as the dihydroxy intermediate in manufacture of terephthalate polymer for polyester fibers, films, and bottles. The second largest use of ethylene oxide is in nonionic surfactants, principally ethylene oxide adducts of linear alcohols to provide biodegradable surfactants for heavy-duty home laundry formulations and anionic ether sulfates for home laundry and dishwashing formulations. Ethoxylated alkylphenols used in industrial and household detergents are also made. Alkanolamines are used in a wide variety of ways, including production of soaps, detergents, and textile chemicals. The glycol ethers have been widely used as industrial coating solvents. The polyglycols are used as intermediates, solvents, humectants, and lubricants.

TABLE 1   Uses of Ethylene Oxide

Ethylene glycol	60 percent
Nonionic surfactants	12
Alkanolamines	8
Diethylene and triethylene glycols	8
Glycol ethers	6
Poly(ethylene glycol)	1

A.   Toxicity

Short-term exposure to ethylene oxide vapor at high concentrations can cause nausea, shortness of breath, central nervous
system depression, and irritation of mucous membranes in humans.   The compound is also described as a protoplasmic poison.
Dilute solutions of ethylene oxide can cause skin blistering, ede-
ma, irritation, and necrosis and eye irritation and necrosis.
Skin contact with liquid ethylene oxide will usually cause burns,
and even short-term skin exposure can lead to skin blisters.
Some acute cases of poisoning have been reported.   Usually,
nausea and vomiting are delayed and, after they occur, there
can be profound weakness of the extremities, convulsive sei-
zures, and secondary lung infection.   A summary of the toxico-
logical effects of ethylene oxide and a referenced summary of
acute effects such as LD_{50} and LC_{50} data are available (1), as
is current information in manufacturers' material safety data
sheets (2).
        Data also indicate that ethylene oxide is a carcinogenic and
mutagenic hazard.   An International Agency for Research on
Cancer (IARC) monograph concludes that there is sufficient
evidence of carcinogenicity of ethylene oxide in animals (3).   Although ethylene oxide causes cancer in mice and rats, evidence
was not yet considered sufficient in 1985 to determine the carcinogenicity of human exposure to ethylene oxide (4).   However,
ethylene oxide should be considered as a potential human carcinogen (5).   The United States Department of Labor's Occupa-
tional Safety and Health Administration (OSHA) has set exposure
limits for workplace atmospheres.   Although one can find relatively high time-weighted averages in the very recent literature,

the OSHA 8-hour, time-weighted average is 1 ppm (5). A 1988 OSHA standard adopted a short-term permissible exposure limit (i.e., an excursion limit) of 5 ppm ethylene oxide averaged over a 15-minute sampling period (6). It is important to use current material safety data sheets when working with ethylene oxide.

The Environmental Protection Agency (EPA) regulates the use and application of mixtures of ethylene oxide and carbon dioxide or halocarbon, which are used as sterilants and fumigants (7), and has established working condition limits and limit levels for allowable ethylene oxide residues in products. Ethylene oxide is also listed as a hazardous air pollutant under the flammable and toxic classes (8) by EPA's Clean Air Act.

There are a number of highly reliable instrument systems available for monitoring ethylene oxide levels in the workplace and the atmosphere. When recommended procedures are closely followed, ethylene oxide does not pose a significant health risk. For example, the use of ethylene oxide as a sterilant in disposable hospital equipment such as catheters and blood-sampling kits and as a grain-storage fumigant has had an enormous cost-effectiveness/benefit ratio. Even though such benefits exist, there have been undesirable effects when plastic articles intended for use in medical practice have been sterilized with mixtures of ethylene oxide and inert compounds (1). The main difficulty centered around sorption of the compound into the plastic and inadequate desorption before use. The transport of gaseous ethylene oxide through linear, low-, medium-, and high-density polyethylene, polypropylene, poly(vinyl chloride), and a fluoroalkylene copolymer, as described by permeability, solubility, and diffusion coefficients, has been reported (9). The measurements were made over a 30°C temperature range using low concentrations of ethylene oxide in helium and a carrier-gas technique (10).

B. Production

Ethylene oxide has been produced by two processes: direct oxidation and the chlorohydrin process. Today, essentially all ethylene oxide in the United States is produced by direct oxidation. Chlorohydrin plants, if still operating, have been converted to production of propylene oxide.

The direct-oxidation process uses a silver catalyst for the vapor-phase oxidation of ethylene with either air or oxygen at about 250°C and 280 psi (1.93 MPa). Carbon dioxide is a by-product.

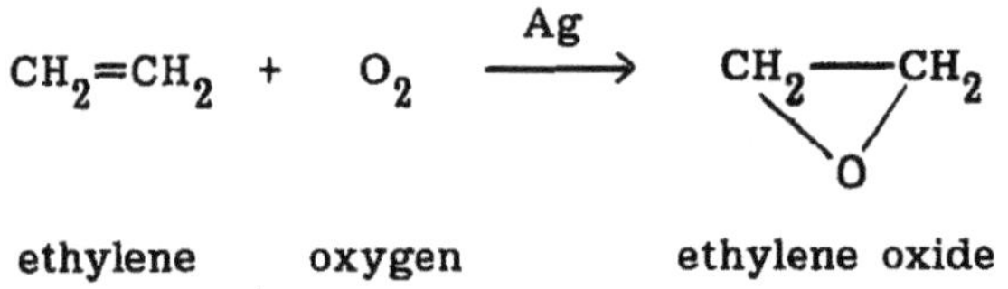

$$CH_2=CH_2 \ + \ O_2 \ \xrightarrow{\text{Ag}} \ CH_2\overset{\displaystyle\diagdown_{\!O}\diagup}{\underline{\qquad}}CH_2$$

ethylene   oxygen   ethylene oxide

The older, chlorohydrin process has a somewhat higher efficiency, but the process consumes chlorine and caustic and produces by-product salt.

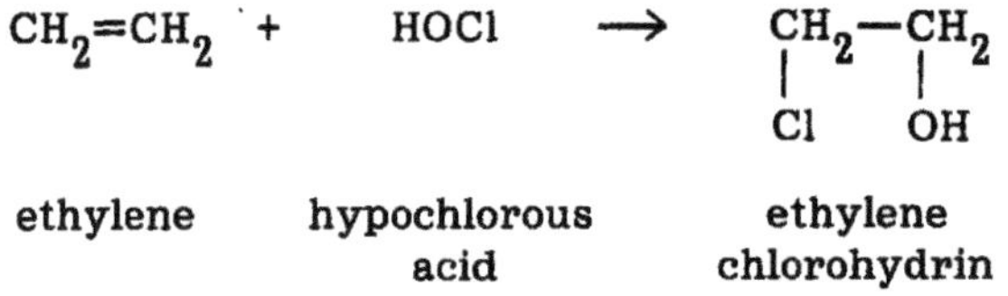

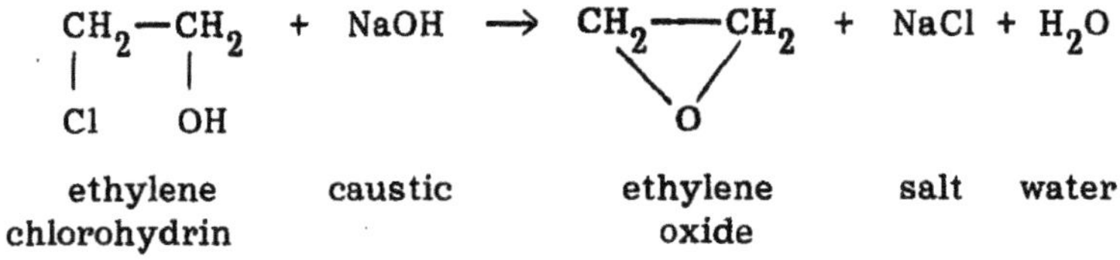

$$CH_2=CH_2 \ + \ HOCl \ \longrightarrow \ \underset{\underset{Cl}{|}}{CH_2}-\underset{\underset{OH}{|}}{CH_2}$$

ethylene   hypochlorous   ethylene
                    acid   chlorohydrin

$$\underset{\underset{Cl}{|}}{CH_2}-\underset{\underset{OH}{|}}{CH_2} \ + \ NaOH \ \longrightarrow \ CH_2\overset{\diagdown_{\!O}\diagup}{\underline{\qquad}}CH_2 \ + \ NaCl \ + \ H_2O$$

  ethylene   caustic   ethylene   salt   water
chlorohydrin   oxide

Ethylene oxide in the United States is produced by 12 companies at 13 locations, as detailed in Table 2. The largest producers are Union Carbide, Shell, Texaco, ICI, BASF, Dow, and Celanese.

Significant ethylene oxide production facilities are operated in Canada by Dow Chemical Company and Union Carbide Corporation affiliates and in Mexico by Petroleos Mexicanos (PEMEX). In Western Europe, ethylene oxide is produced by a number of chemical manufacturers including BASF, British Petroleum, Bayer, Hoechst, Huels, Dow, Shell, and ICI. Producers in Japan include Mitsubishi Petrochemical, Mitsui Petrochemical Industries and Mitsui Toatsu, Nippon Shokubai Kagaku Kogyo, Nisso Maruzen, and Nisso Petrochemicals Industries.

Ethylene oxide is a colorless gas under normal ambient conditions. It is odorless at low levels and cannot be detected by smell until the concentration is very high—above safe exposure levels. Above about 500 ppm, ethylene oxide has an etherlike odor. Selected physical characteristics of ethylene oxide are given in Table 3.

TABLE 2   Ethylene Oxide Production in the United States

Producing company	Approximate annual production capacity	
	Million pounds	Metric tons $\times 10^{-3}$
BASF Wyandotte	500	227
Dow Chemical	470	214
Hoeschst-Celanese	450	205
ICI	500	227
Norchem	220	100
Olin	130	59
PD Glycol	460	209
Shell Chemical	800	364
SunOlin	100	45
Texaco	700	318
Texas Eastman	200	91
Union Carbide	2000	909

Up-to-date details regarding the handling of ethylene oxide and other alkylene oxides are available from manufacturers. A code of practice for the safe storage and handling of ethylene oxide provides a great deal of useful information (11). Its low flash point of −17.8°C and its flammability in air at concentrations of 3 volume percent and greater, as well as the possibility of explosion at high concentrations, should be borne in mind and suitable precautions taken when using the chemical. Ethylene oxide is very liable to polymerization that is initiated by acids, bases, anhydrous metal chlorides, and metal oxides. Metals such as copper and copper alloys, as well as other acetylide-forming metals, should not be in contact with ethylene oxide. Polymerization is also initiated by iron rust, and rust should be removed from equipment containing ethylene oxide. The polymerization is very exothermic and, if not controlled after initia-

TABLE 3 Selected Properties of Alkylene Oxides

Property	Ethylene oxide	Propylene oxide	1,2-Butylene oxide	α-Epichlorohydrin
Autoignition temp., °C	428.9	465.0	—	—
Boiling point (1 atm), °C	10.4	34.5	63.3	116.5
Density, g/cc (at °C)	0.88267 (10.7)	0.82326 (25.0)	0.82993 (20.0)	1.1801 (20.0)
Flammability levels in air, volume percent	3–100	2.1–21.5	1.5–18.3	—
Freezing point, °C	−111.7	−111.9	−150*	−57
Heat of combustion (25°C), Btu/lb	12,344	~13,800	14,665	—
Heat of vaporization (1 atm), Btu/lb	250	201	177	—
Molecular weight	44.05	58.08	72.11	92.53

Refractive index, nD, 20°C	1.3634†	1.3657	1.3840	1.438
Vapor pressure (20°C), mmHg	1095	440	141	13

*Not a freezing point. Compound sets to a glass at this temperature, and it represents a vitrefication temperature.

†Value is at 0°C.

Synonyms

Ethylene oxide: dimethylene oxide, dihydrooxirane, oxirane, 1,2-epoxyethane, oxacyclopropane, etileno oxido (Spanish), aethyleneoxide (German), oxyde d'ethylene (French), ossido dietiline (Italian), oxiraan (Dutch)

Propylene oxide: 1,2-epoxypropane, 1,2-propylene oxide, methyl oxirane, methyl ethylene oxide, propileno oxido (Spanish), oxyde de propylene (French)

1,2-butylene oxide: 1,2-epoxybutane, ethyl oxirane, ethyl ethylene oxide, butileno oxido (Spanish), 1,2-oxyde de butylene (French)

Epichlorohydrin: alpha-epichlorohydrin, 1-chloro-2,3-epoxy propane, glycidyl chloride, gamma-chloropropylene oxide

tion, is self-accelerating; explosive decomposition can take place. Factors such as these make proper handling an important safety factor when dealing with ethylene oxide, as well as with other alkylene oxides.

C. Uses

The largest single use of ethylene oxide is in the production of ethylene glycol by direct, noncatalytic, liquid-phase hydration.

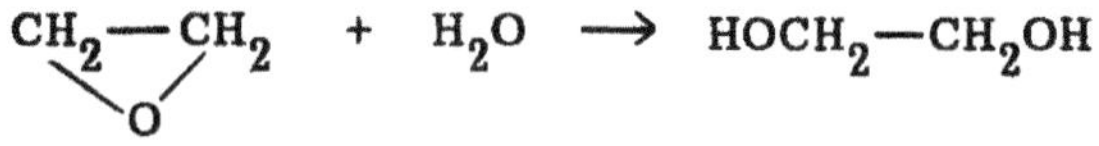

$$CH_2 \!\!-\!\! CH_2 \;+\; H_2O \;\longrightarrow\; HOCH_2 \!\!-\!\! CH_2OH$$
$$\diagdown O \diagup$$

    ethylene oxide   water   ethylene glycol

The largest outlet for ethylene glycol, accounting for about half the total production in the United States, is in antifreeze formulations. This production is principally for automotive use, but includes deicing fluids; for example, those used on aircraft. The second largest use is for the production of polyesters such as poly(ethylene terephthalate) for fibers, films (decorative, packaging, and photographic), and containers (bottles for soft drinks, mouth wash, patent medicines, etc.), and of ethylene glycol adipates for polyurethane elastomers.

    Diethylene glycol and triethylene glycol are obtained principally as by-products of ethylene glycol manufacture.

$HO-CH_2-CH_2-O-CH_2-CH_2-OH$ $HO-CH_2-CH_2-O-CH_2-CH_2-O-CH_2-CH_2-OH$

       diethylene glycol   triethylene glycol

Diethylene glycol is reacted with adipic acid to manufacture polyester oligomers that are used in the manufacture of polyurethanes for various end uses. Both compounds are used in the manufacture of alkyds and as reactive diluents and crosslinking agents in coatings.

    Poly(ethylene glycol)s are made by reaction of ethylene oxide with water, ethylene glycol, or diethylene or triethylene glycol using sodium or potassium hydroxide catalyst.

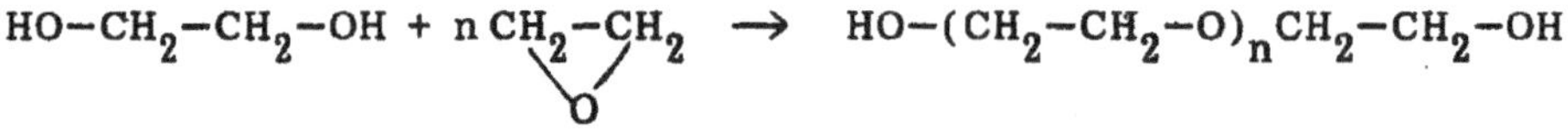

$$HO-CH_2-CH_2-OH + n\,CH_2\!\!-\!\!CH_2 \;\longrightarrow\; HO-(CH_2-CH_2-O)_n CH_2-CH_2-OH$$
$$\diagdown O \diagup$$

      ethylene   ethylene   poly(ethylene glycol)
       glycol   oxide

These processes involve the simplest of the polymerization reactions of ethylene oxide, leading, in these cases, to polymers of relatively low molecular weight that are used in surfactants, functional fluids (brake fluids and hydraulic fluids, lubricants), cosmetics, emollients, and vinyl plasticizers. The value of n can range from about 5 (although lower molecular weight oligomers exist, as indicated above) to 500 for compounds termed poly(ethylene glycol).

The second largest use of ethylene oxide is in the production of nonionic and anionic surfactants. These fall into two broad classes. One is ethoxylates of alkylphenols, principally nonyl phenol and, to a lesser extent, dodecyl phenol. These

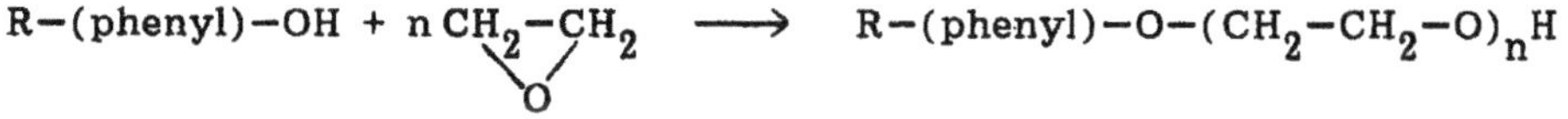

$$R-(phenyl)-OH + n\ CH_2-CH_2 \longrightarrow R-(phenyl)-O-(CH_2-CH_2-O)_nH$$
$$O$$

alkylphenol   ethylene oxide   alkylphenyl-poly(oxy-
                                                 ethylene glycol)

ethoxylates of alkyl phenols are used chiefly as industrial detergents or are converted by sulfating or phosphating to anionic (ester) surfactants. Use of these alkylphenol ethoxylates is slowly being restricted because they are not readily biodegradable.

The second class includes polyethylene glycol alkyl esters, ethoxylated linear aliphatic alcohols, ethoxylated natural fatty acids and oils, and alkanolamides. These products, together with ethylene oxide block copolymers with propylene oxide, form a class of ethoxylate surfactants that are biodegradable. Linear alcohol ethoxylates find extensive use in heavy-duty laundry detergents.

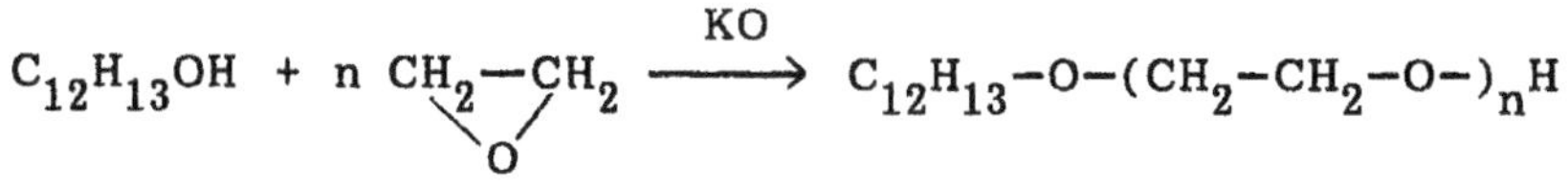

$$C_{12}H_{13}OH + n\ CH_2-CH_2 \xrightarrow{KO} C_{12}H_{13}-O-(CH_2-CH_2-O-)_nH$$
$$O$$

lauryl   ethylene   lauryl-poly(oxyethylene
alcohol   oxide   glycol)

These products are particularly effective detergents for synthetic fabrics and fabric blends at cool washing temperatures. Sulfates of the linear alcohol ethoxylates are used in home laundry and dishwasher formulations.

II. PROPYLENE OXIDE

World annual production of propylene oxide is about 4 billion
pounds (1.8 million metric tons) (1982). About 42 percent of
this production is in the United States, about 44 percent is in
Western Europe, and about 10 percent is in Japan. The principal uses for propylene oxide in the United States are listed in
Table 4.

The polyols used for production of polyurethanes are the
reaction products of propylene oxide, often in copolymerization
with a lesser amount of ethylene oxide, and polyhydroxy compounds such as glycerol, trimethylolpropane, and sorbitol or
polyamines of which ethylenediamine or 2,6-diaminotoluene are
examples. These polyols usually have a functionality—that is,
an average number of reactive hydroxyl groups per molecule—
greater than 2, and usually greater than 3. Depending on use,
these polyols have molecular weights ranging from a few hundred to 5000 to 10,000. The polyurethane polyols are intermediates used to make polyurethane foams and elastomers by further
reaction with polyisocyanates. About 72 percent of the polyurethane polyols are consumed in the production of flexible urethane foams that are used extensively in furniture and upholstery cushioning, mattresses, automotive seating, sound-deaden-
ing applications, home carpet underlayment, and protective
packaging. About 8 percent is employed in the manufacture of
rigid foams used as thermal insulation in construction and appliances. Twenty percent is used in the manufacture of urethane
elastomers. Urethane elastomers include large moldings made by
reaction injection molding (RIM) processes for automobile exterior parts (fascias), bumper systems, and machine housings.

The second largest use of propylene oxide is for propylene
glycol. Propylene glycol,

$$HO-CH_2-CH-OH$$
$$|$$
$$CH_3$$

is used in the manufacture of unsaturated polyester resins used
in reinforced plastics. In Europe, this glycol is used as an antifreeze because its animal toxicology is much better than that
of ethylene glycol.

TABLE 4   Uses of Propylene Oxide

Polyurethane polyols	64 percent
Propylene glycol	19
Dipropylene glycol	4
Glycol ethers	3

A. Toxicity

Propylene oxide is a skin and eye irritant, a mild central nervous system depressant, and a mild protoplasmic poison. Skin contact causes necrosis and skin irritation. Excessive exposure to vapors may cause eye, lung, and respiratory tract irritation and central nervous system effects that may be characterized by general depression, coordination difficulties, and ataxia. A secondary effect of excessive exposure is pulmonary infection. Summaries of acute responses of dogs, mice, rats, and guinea pigs to propylene oxide exposure, as well as pathological findings, have been tabulated (12). Propylene oxide has a sweet, alcoholic odor that is detectable at a concentration of about 200 ppm. The detectable odor level is above the suitable repeated and prolonged exposure level. Current information is available from material safety data sheets (13).

B. Production

There are two major processes used to produce propylene oxide: the chlorohydrin process and peroxidation of propylene. More than half of world production is by the chlorohydrin route. In this process, the first step is reaction of propylene with hypochlorous acid to obtain propylene chlorohydrin.

$$CH_3CH=CH_2 \quad + \quad HOCl \quad \longrightarrow \quad CH_3CH-CH_2$$
$$\underset{\text{OH}\quad\text{Cl}}{}$$

propylene	hypochlorous acid	propylene chlorohydrin

Propylene chlorohydrin is then treated with caustic or calcium hydroxide to obtain propylene oxide.

$$CH_3-CH-CH_2 + NaOH \rightarrow CH_3CH-CH_2 + NaCl + H_2O$$

| | | | |
| OH Cl | | O | |

propylene   caustic   propylene   salt   water
chlorohydrin   oxide

While the chlorohydrin process is efficient, it consumes caustic and chlorine and produces by-product salt. It is difficult to operate this process unless production is associated with a sizable chlor-alkali business. In the United States, about half of the production capacity for propylene oxide uses the chlorohydrin process.

In the peroxidation processes, propylene oxide is obtained by reaction of propylene with either a hydroperoxide or with peracetic acid. In the United States, manufacture is limited to oxidation of propylene by either of two hydroperoxides: t-butyl hydroperoxide or ethylbenzene hydroperoxide. A hydrocarbon, either isobutane or ethylbenzene, is first oxidized to the corresponding hydroperoxide and then the hydroperoxide is used to oxidize propylene to give propylene oxide with a useful alcohol by-product.

$$(CH_3)_3CH + O_2 \longrightarrow (CH_3)_3COOH$$

t-butane   oxygen   t-butyl hydroperoxide

$$CH_3CH=CH_2 + (CH_3)_3COOH \rightarrow CH_3CH-CH_2 + (CH_3)_3COH$$

propylene   t-butyl   propylene   t-butanol
                  hydroperoxide   oxide

$$C_6H_5CH_2-CH_3 + O_2 \rightarrow C_6H_5CHOOH$$
$$CH_3$$

ethylbenzene   oxygen   ethylbenzene
                                              hydroperoxide

$$CH_3CH=CH_2 + C_6H_5CHOOH \rightarrow CH_3CH-CH_2 + C_6H_5CHOH$$
$$CH_3 \qquad\qquad O \qquad\qquad CH_3$$

propylene   ethylbenzene   propylene   phenyl methyl
                hydroperoxide   oxide   carbinol

Isobutanol can be dehydrated to isobutene and used as an intermediate or converted to methyl t-butyl ether, used in unleaded gasoline. Phenyl methyl carbinol can be dehydrated to styrene, and it is a major intermediate for this monomer.

Propylene oxide is manufactured in the United States principally by Dow Chemical Company (chlorohydrin process) and ARCO Chemical Company, a Division of Atlantic Richfield (the peroxidation process based on both isobutane and ethylbenzene). In Western Europe, major producers include Dow Chemical Company, ARCO Chemical, Shell, BASF, and Bayer, through a subsidiary. For a time, a peracetic acid process for propylene oxide was operated in Japan; however, competitive sources of byproduct acetic acid forced discontinuation of this process.

Peroxidation with peracetic acid is used to produce a number of other epoxides that are commercially available. These include epoxidized soybean oil and linseed oil used as vinyl plasticizers and hydrochloric acid scavengers in processing aliphatic halogen-containing polymers and cycloaliphatic diepoxides that are used in epoxy polymers for electrical and electronic applications. This process has also been used to make a large number of alkylene epoxides, the polymers of which have been described in the literature, but which have never reached commercial status.

The properties of propylene oxide are detailed in Table 3. In general, this compound should be handled in the same manner as ethylene oxide.

C. Uses

The largest use of propylene oxide is in the manufacture of polyols for polyurethane production. These propylene oxide polyols are propoxylates of low-molecular-weight polyhydroxy compounds such as glycerine or some sugars or polyamines. The propoxylation is catalyzed by caustic, usually potassium hydroxide, and the resulting polyols have molecular weights from a few hundred to 5000 or 10,000 and have hydroxyl functionalities (average number of hydroxyl end groups per molecule) of more than 2 and usually of 3 to 6. The lower molecular weight polyols are used to make rigid foams used principally as thermal insulation in construction and refrigeration. Higher molecular weight polyols are used for their flexibility and load-bearing properties in furniture cushioning and automobile and truck seating, as well as in public transportation, bedding, and home carpet underlayment. These polyols are also used in the manu-

facture of urethane elastomers and in large moldings for automobile exterior parts (e.g., fascias), sports equipment (snowmobile bodies), and tractor housings.

Propylene glycol is made by liquid-phase reaction of water
with propylene oxide at about 200°C.

$$CH_3-CH-CH_2 \quad + \quad H_2O \quad \rightarrow \quad CH_3-CH-CH_2-OH$$

$$\underset{O}{\diagdown\diagup} \qquad\qquad\qquad\qquad\qquad \underset{OH}{|}$$

| propylene | water | 1,2-propylene |
| oxide | | glycol |

The largest use of propylene glycol is in the production of unsaturated polyesters used in reinforced plastics. The unsaturated polyester is usually the polycondensation product of
phthalic anhydride, with some maleic anhydride, and propylene
glycol. This unsaturated polyester is then mixed with fiberglass, styrene monomer, and a free-radical catalyst and molded
under pressure. During the molding, the styrene copolymerizes
with the polyester unsaturation to give a fiber-reinforced thermoset plastic that may find use as construction panels, diving
boards, automobile exteriors, pleasure boat hulls, etc.

Propylene glycol has a very sizable usage because of its
low toxicity. Unlike ethylene glycol, which is oxidized to oxalic
acid, the oxidation products of propylene glycol form relatively
soluble salts that are not harmful in quantities that are usually
encountered. Propylene glycol is used widely in food, cosmetics, and drugs. In foods, it is used as a flavoring solvent and
as a humectant in bread and baked goods; it is also used as an
ink vehicle in packaging and as a packaging lubricant. It is
also employed as a tobacco humectant and as a cosolvent in
water-based surface coatings.

Dipropylene glycol and tripropylene glycol, which are to
some extent made along with propylene glycol during the hydration of propylene glycol, are also made specifically by reaction
of propylene glycol with propylene oxide.

$$HO-CH-CH_2OH + CH_2-CH-CH_3 \rightarrow HO-CH-CH_2-O-CH_2-CH-OH$$

$$\underset{CH_3}{|} \qquad\qquad \underset{O}{\diagdown\diagup} \qquad\qquad\quad \underset{CH_3}{|} \qquad\qquad\qquad \underset{CH_3}{|}$$

| propylene | propylene | dipropylene glycol |
| glycol | oxide | |

Dipropylene glycol and tripropylene glycol are also used in making unsaturated polyesters to increase the polyester flexibility. They are also esterified to form a variety of vinyl plasticizers and are used in rubber, paint, and adhesive formulations.

A series of solvents is manufactured by reaction of alcohols, ROH, with propylene oxide to form polypropylene glycol ethers,

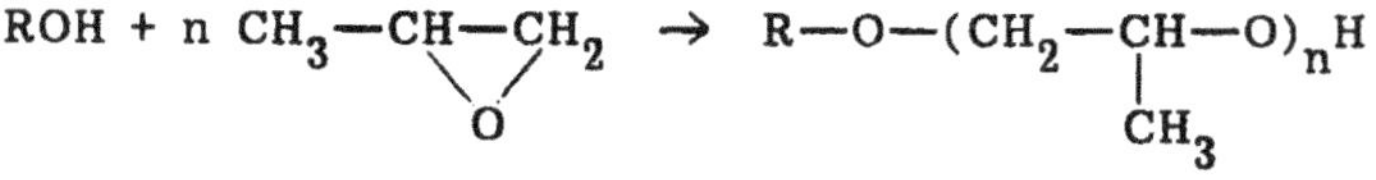

$$ROH + n\ CH_3{-}CH{-}CH_2 \rightarrow R{-}O{-}(CH_2{-}CH{-}O)_n H$$

where R may be methyl, butyl, or allyl, for example, and n may be 2 or 3. This reaction and that for higher molecular weight polyethers can be catalyzed with a caustic, usually potassium hydroxide. In addition, ethylene oxide can be used as a comonomer for both block and random copolymers, which find wide use as hydraulic fluids, textile lubricants, metal-working lubricants, and surfactants. These poly(alkylene oxide)s will be discussed in detail in later chapters of this book.

III. BUTYLENE OXIDE (1,2-EPOXYBUTANE)

Butylene oxide, or 1,2-epoxybutane, is produced in the United States by Dow Chemical Company. The current production is about 8 million pounds (3,640 metric tons) per year by a chlorohydrin process similar to that used to make propylene oxide. It is available as 1,2-butylene oxide or as an 80-90/10-20 mixture of 1,2- and 2,3-butylene oxide. In Europe, 1,2-butylene oxide is manufactured principally by BASF.

$$C_2H_5{-}CH{=}CH_2\ +\quad HOCl\quad \rightarrow\quad C_2H_5{-}CH{-}CH_2$$

butene-1	hypochlorous	butene
	acid	chlorohydrin

$$C_2H_5{-}CH{-}CH_2 + caustic \rightarrow C_2H_5{-}CH{-}CH_2 + salt$$

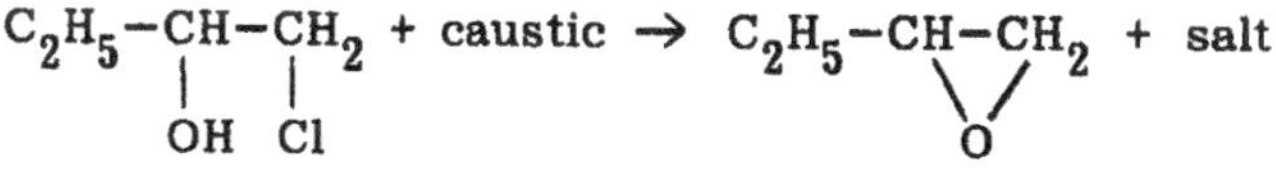

butene
chlorohydrin

1,2-epoxybutane

The properties of butylene oxide are listed in Table 3.

Butylene oxide is used principally as an HCl acceptor in stabilizer formulations for 1,1,1-trichloroethane, a dry-cleaning and metal-cleaning solvent. It is also used as a pharmaceutical intermediate and as a comonomer in making nonionic surfactants and copoly(alkylene oxide)s. Other uses include preparation of gasoline additives, butanol amines, butylene glycols, and their ether and ester derivatives.

A. Toxicity

Butylene oxide has a sweetish, but very disagreeable, odor. The liquid is markedly irritating to eyes and skin. While the vapors are moderately anesthetic, the concentration required for this effect is very high. Even though the odor is disagreeable, odor detection is no safeguard for health and safety (14). Inhalation may be fatal. The compound is extremely destructive to the upper respiratory tract, mucous membranes, skin, and eyes. Oral LD_{50} was found to be 500 mg/kg when tested with rats, and skin LD_{50} was 2100 mg/kg when tested with rabbits (15). The compound is listed as a carcinogen.

IV. EPICHLOROHYDRIN (1-CHLORO-2,3-EPOXY PROPANE)

World production of epichlorohydrin is about 1 billion pounds (455,000 metric tons) per year. Of this production, just slightly less than half is made in the United States, and about 40 percent is produced in Western Europe. In the United States and Europe, the major producers are Dow Chemical Company and Shell. Several companies in Japan produce about 100 million pounds (45,500 metric tons) of epichlorohydrin per year.

A. Toxicity

Epichlorohydrin has been listed as a possible carcinogen and mutagen. Worker exposure is limited to 5 ppm in the atmosphere (1987). The compound is a low-viscosity, colorless liquid that boils at 116°C and freezes at −57.2°C. At 20°C, it has a vapor pressure of 13 mm mercury. Epichlorohydrin has a chloroformlike odor that is very irritating to eyes and skin. Acute toxicity data as well as other animal and human responses have been summarized (16).

B. Production

Epichlorohydrin is made by a three-step process. First, propylene is chlorinated to form allyl chloride, which is then reacted with hypochlorous acid to give glycerol dichlorohydrin. Caustic treatment of glycerol dichlorohydrin then produces epichlorohydrin.

$$CH_2\!=\!CH\!-\!CH_3 \quad + \quad Cl_2 \quad \longrightarrow \quad CH_2\!=\!CH\!-\!CH_2Cl \quad + \quad HCl$$

  propylene   chlorine   allyl chloride

$$CH_2\!=\!CH\!-\!CH_2Cl \quad + \quad HOCl \quad \longrightarrow \quad CH_2Cl\!-\!\underset{\underset{OH}{|}}{CH}\!-\!CH_2Cl$$

and

$$CH_2Cl\!-\!\underset{\underset{Cl}{|}}{CH}\!-\!CH_2OH$$

  allyl chloride   hypochlorous   symmetrical and
                             acid   unsymmetrical glycerol
                                                   chlorohydrin

symmetrical or   +   NaOH $\longrightarrow$ $CH_2Cl\!-\!CH\!-\!CH_2$ + NaCl
unsymmetrical $\diagdown\!O\!\diagup$
glycerol chlorohydrin

epichlorohydrin

In Japan, a process has been used in which allyl chloride is epoxidized with hydrogen peroxide to give epichlorohydrin directly.

C. Uses

At one time, the major use of epichlorohydrin was as an intermediate in the manufacture of glycerol. Now, 55 percent of the production is consumed in the making of epoxy resins. The major epoxy component made is the diglycidylether of bisphenol A.

Specialty elastomers based on epichlorohydrin have been produced by both Hercules and BF Goodrich. Elastomers have been made by copolymerization of epichlorohydrin with ethylene oxide and by terpolymerization of epichlorohydrin with ethylene

oxide and propylene oxide. In 1984, BF Goodrich disclosed a
method of polymerizing epichlorohydrin to make a specialty elastomer, HYDRIN^{TM} 100, via a unique cationic polymerization
process. Copolymers and terpolymers are also marketed.

V. HIGHER 1,2-EPOXIDES

1,2-Epoxides of higher olefins are available as specialty chemicals. They can be prepared either by dehydrohalogenation of
the corresponding halohydrin or by epoxidation of the corresponding olefin with an organic peracid such as peracetic acid
or perbenzoic acid (17,18). A large number of these epoxides
have been prepared (19) and are described in the literature.

A related class of epoxides is the omega-epoxyalkanoic esters. These 1,2-epoxides can also be prepared from the corresponding terminal alkenoate ester by peroxidation and have been
polymerized by Vogl (20).

The polymerization of the higher olefin oxides proceeds by
reactions essentially the same as those that will be discussed for
propylene oxide and butylene oxide. Many of these have been
described by Vandenberg (21). Health characteristics and animal and human responses can vary markedly from one olefin
epoxide to another (22).

REFERENCES

1. G. D. Clayton and F. E. Clayton, Patty's Hygiene and
    Toxicology, V2A Toxicology, John Wiley & Sons, New York,
    1981, pp. 2166–2186.
2. MSDS, Union Carbide Corp., Ethylene Oxide, May 1987.
3. IARC Monograph, Evaluation of the Carcinogenic Risk of
    Chemicals to Humans. 36, IARC, Lyon, France, 1984.
4. U.S. Department of Health and Human Services, Fourth
    Annual Report on Carcinogens, NTP 85-002, Public Health
    Service, 1985.
5. OSHA, 29 CFR 1910.1047.
6. OSHA, 53 Federal Register 11414 (4/6/88); J. Coatings
    Tech. 60 (7):20 (1988).
7. G. D. Clayton and F. E. Clayton, Patty's Hygiene and
    Toxicology, V2B, Toxicology, John Wiley & Sons, New
    York, 1981, p. 3109.

8. K. M. Kearney, Semiconductor Internat.:78–83 (October 1988).
9. A. Phatak, C. M. Burns, and R. Y. N. Huang, J. Appl. Polymer Sci. 34:1835 (1987).
10. T. L. Caskey, Mod. Plastics 45:148 (1967).
11. Imperial Chemical Ind. Ltd, Shell Chemicals UK Ltd., and Union Carbide UK Ltd., The Handling and Storage of Ethylene Oxide, CPP4, The Chemical Industries Safety & Health Council of the Chemical Industries Association, London, August 1975.
12. G. D. Clayton and F. E. Clayton, Patty's Hygiene and Toxicology, V2A Toxicology, John Wiley & Sons, New York, 1981, pp. 2186–2191.
13. MSDS, Dow Chemical USA, Propylene Oxide, April 18, 1989.
14. G. D. Clayton and F. E. Clayton, Patty's Hygiene and Toxicology, V2A Toxicology, John Wiley & Sons, New York, 1981, pp. 2162–2165.
15. Aldrich Chemical Co., Material Safety Data Sheet, 1,2-Epoxybutane, 99% (3/18/87).
16. G. D. Clayton and F. E. Clayton, Patty's Hygiene and Toxicology, V2A Toxicology, John Wiley & Sons, New York, 1981, pp. 2242–2247.
17. S. Winstein and R. B. Henderson, in Heterocyclic Compounds (R. C. Elderfield, ed.), John Wiley & Sons, New York, 1950, p. V1.
18. D. Swern, in Organic Reactions, Vol. VII (R. Adams et al. eds.), John Wiley & Sons, New York, 1953.
19. L. E. St. Pierre, in Polyethers (N. G. Gaylord, ed.), Interscience Publishers, New York, 1963, p. 83.
20. O. Vogl, J. Muggee, and D. Bansleben, Polymer J. 12 (9):677 (1980).
21. E. J. Vandenberg, J. Polymer Sci., Part A1, 7:525 (1969).
22. B. L. Van Duuren, L. Langseth, B. M. Goldschmidt, and L. Orris, J. Nat. Cancer Inst. 39 (6):1217 (Dec. 1967).

BIBLIOGRAPHY

S. A. Cogswell, "Ethylene Oxide, Marketing Research Report," in Chemical Economics Handbook, SRI International, Menlo Park, 1986.

A. O. Masilungen, "Propylene Oxide, Marketing Report," in Chemical Economics Handbook, SRI International, Menlo Park, 1984.

Glycols (G. O. Curme, ed.), Reinhold Publishing Company,
   ACS Monograph Series, New York, 1952.

Y. Okamoto, Polymer Preprints 25:264 (1984); S. Yu, Polymer
   Preprints 25:117 (1984).

K. Wheeler, T. Ruess, and S. Takahashi, "Epichlorohydrin,
   Marketing Research Report," in Chemical Economics Handbook, SRI International, Menlo Park, 1987.

3

Early History of the Poly(alkylene oxide)s

The history of the poly(alkylene oxide)s predates modern structural organic chemistry. The first references, however, presage both structural organic and polymer chemistry. In 1859, Lourenço (1) reported that he had heated ethylene glycol and ethylene dibromide in sealed tubes at 115°C to 120°C and isolated, by distillation, a series of materials that were indistinguishable by elemental analysis. He reported his products using the compositional notation of the day:

    C2H4)
    O1   )n

His paper indicates that he had made and isolated products ranging from diethylene glycol to hexaethylene glycol, in modern notation:

$$HO(C_2H_4O)_nH \quad n = 1-6$$

At about the same time, Wurtz (2) isolated products from the reaction of ethylene oxide with water, ethylene glycol, and acetic acid that clearly were polyethylene glycols and polyethylene glycol acetate. Wurtz also observed a solid, crystalline product melting at 56°C and soluble in water when ethylene oxide stood at room temperature in contact with a trace of alkali or zinc chloride. Wurtz also reported (3) that when ethylene oxide was placed in a sealed tube and stored for about a year, it was converted into a white solid and that zinc chloride and caustic accelerated the reaction. Later, Roithner (4) found that the reaction proceeded more rapidly at higher temperatures (about 60°C)

In 1929, H. Staudinger and O. Schweitzer (5) studied the polymerization of ethylene oxide in contact with a number of catalysts and separated the products obtained into components differing by molecular weight. They reported isolating the pure hexamer boiling at 325°C at 0.025 mmHg. Later, Staudinger and Lohman (6) studied the rate of polymerization of ethylene oxide with a number of catalysts; the results are summarized in Table 1.

Samples of polymers prepared by Staudinger and Lohman were analyzed by The Svedberg, using his newly invented ultracentrifuge, confirming molecular weights as high as several hundred thousand. This collaboration of Staudinger and Svedberg in characterizing samples of poly(ethylene oxide), as well as samples of polystyrene prepared at about the same time, confirmed the macromolecular hypothesis, which established polymer chemistry during the 1920s.

TABLE 1 Early Work on the Rate of Polymerization of Ethylene Oxide at 20°C

Catalyst	Polymer yield, percent	Molecular weight	Time of reaction
Trimethylamine)			
Na or K)	5	2000	1−2 weeks
$SnCl_4$)			
Sodamide	1−2	10,000	2−3 months
ZnO	10−20	60,000	2−3 months
SrO	10−20	100,000	3−4 months
CaO	50	120,000	2 years

In 1929, workers at the Indigo Laboratory of I. G. Farben-industrie in Ludwigshafen reported (7) that Dr. Staudinger had pointed out to them that ethylene oxide in the presence of small amounts of catalysts such as potassium, tin tetrachloride, and trimethylamine can occasionally polymerize completely in a few seconds at ordinary temperatures. Since the large heats of re-action caused explosive destruction of the reaction tubes, and because ethylene oxide was used in large amounts at I. G. Far-benindustrie, an investigation of ethylene oxide in the presence of different materials, and particularly iron and iron compounds, was undertaken. The researchers found that explosions could result from a number of materials when they were added to ethylene oxide. Other products were more mild in their reaction characteristics. Table 2 contains qualitative results and heats of reaction from a number of experiments.

The I. G. Farbenindustrie investigators concluded that highly divided iron oxide of the voluminous type that is obtained from iron carbonyl was the most effective iron compound for converting ethylene oxide to other substances. When such iron oxide (10 percent) was heated with ethylene oxide for 36 hours at 98°C in a steel bomb, the entire charge formed a brown pol-ymeric mass. Although 10 percent concentrations of iron oxide and of aluminum oxide resulted in polymerization, when 20 per-cent of these compounds were used, violent, explosive decompo-sition resulted. In one series of experiments, iron rust from inside an ethylene oxide cylinder was powdered, sieved, and used as a catalyst for ethylene oxide polymerization in a sealed steel bomb at 95—98°C. The results, shown in Table 3, indi-cated that the extent of the relatively slow polymerization in-creased with catalyst concentration and reaction time. It is in-teresting to note that at this time it was thought that ethylene oxide polymerized by means of a rearrangement mechanism where-in acetaldehyde was first formed and then polymerized. How-ever, it was felt possible, but not very probable, that ethylene oxide could directly polymerize without undergoing rearrange-ment.

Roithner (4) had concluded that poly(ethylene oxide) had a cyclic structure, essentially dioxane for the dimer and then larger ring structures for the higher polymers. Some later workers concluded that while lower polymers were cyclic, higher polymers were probably linear. Much of this confusion was due to a lack of understanding, early on, of the several mechanisms of polymerization, and no differentiation was made between acid

TABLE 2   Results of Contacting Ethylene Oxide and Various
Materials (From Ref. 7.)

Material	Heat of reaction, kcal/kg	Result
Potassium	—	Explosive reaction
Powdered aluminum	—	Very slow polymerization
Powdered copper	—	Very slow polymerization
Aluminum chloride	+680	Very violent reaction
Ferric chloride	+252	Violent reaction
Stannic chloride	+371	Violent reaction
Phosphorous pentachloride	—	Rapid reaction
Titanium tetrachloride	—	Violent reaction
Magnesium chloride	+89	Not determined
Zinc chloride	−27	Slow polymerization
Stannous chloride	−49	Slow polymerization
Cuprous chloride	−79	Slight polymerization
Iron oxide (10% level)	—	Polymerization
Iron oxide (20% level)	—	Explosive decomposition
Aluminum oxide (10% level)	—	Polymerization
Aluminum oxide (20% level)	—	Explosive decomposition

catalysis and base catalysis.   It was true that acid catalysis did
produce a sizable product, dioxane, and some cyclic tetramer.
Further, acid catalysis of propylene oxide did produce cyclic
structures.   Wurtz (8) found that both glacial and aqueous
acetic acid with ethylene oxide mainly formed the diacetate of
ethylene glycol.

During the 1930s, poly(ethylene glycol)s were made commercially by base initiation of the addition of ethylene oxide to

TABLE 3 Polymerization of Ethylene Oxide
at 95—98°C in the Presence of Rust from an
Ethylene Oxide Cylinder (From Ref. 7.)

Percent rust	Reaction time, hr	Percent polymer formed
10	40	10.4
10	75	16.0
20	16	10.8
20	42	25.5
40	72	33.3

ethylene glycol and diethylene glycol. Perry and Hibbert (9)
proposed a stepwise addition mechanism for this polymerization
based on the demonstration that the product was a glycol with
two end-group hydroxyls (10). Finally, in 1940, Paul Flory
(11) described the kinetics of the base-initiated polymerization
of the cyclic ethers such as ethylene oxide and showed that the
kinetics predicted a narrow distribution of polymer molecular
weights. Experiments verified this prediction of a Poisson distribution of molecular weights. This distribution predicts that,
as the average polymer molecular weight becomes large, the ratio of the weight to number average molecular weight approaches
unity, provided certain reaction conditions are met:

All reaction chains are initiated simultaneously.
Monomer reacts only with active sites on the polymer chain.
There is no reaction-chain termination other than total depletion
    of monomer.

During this time, a number of applications for the polyethylene
glycols, which are still in use, were developed, including use
in pharmaceuticals and cosmetics; plastics, rubber-processing
aids, and mold-release agents; textile lubricants and dye dispersants; metal working, polishing, and electroplating; and as
a binder for synthetic detergents.
    One of the earliest attempts to polymerize propylene oxide
was reported by Levene and Walti (12) in 1927. A small amount

of propylene oxide was heated in a sealed tube for several weeks at 165°C. A mixture of products was obtained containing some dimethyldioxane and some presumably linear polyethers, including di-, tri-, and tetrapropylene glycol. Staudinger used stannic chloride to polymerize propylene oxide to a mixture of products. During the 1930s, considerable work was published by Meerwein concerning the polymerization of propylene oxide to cyclic products, and a number of patents were issued to I. G. Farbenindustrie, AG (13).

It was during the 1940s that the base-initiated polymerization of propylene oxide began to be used to produce products that are still in use today (14). In a series of patents by H. R. Rife and F. H. Roberts (15), poly(propylene oxide)s are described that are initiated in the presence of alcohols to produce ether propoxylates. These liquid polymers had methoxy or butoxy groups on one end of the polyether chain and were used as lubricants and hydraulic fluids. This work was quickly followed by other derivatives, which included esters and mixed alkoxylates of propylene oxide and ethylene oxide (16). Uses included antifoams and emulsifiers, coating solvents, ceramic glazes and binders, and synthetic lubricants for internal combustion engines (17).

REFERENCES

1. A. Lourenço, Compt. rend. 49:619 (1859).
2. A. Wurtz, Compt. rend. 49:813 (1859); Ann. chim. et phys. 69 (3):317 (1863); Ber. 10:90 (1877).
3. A. Wurtz, Compt. rend. 83:1141 (1877); Compt. rend. 86:1176 (1878).
4. E. Roithner, Monatash 15:665 (1894).
5. H. Staudinger and O. Schweitzer, Ber. 62:2375 (1929).
6. H. Staudinger and H. Lohman, Ann. chim. 505:41 (1933).
7. I. G. Farbenindustrie A. G. Report, Reactivity of Ethylene Oxide, (9/6/29) in Field Information Agency Technical Report, The Manufacture of Ethylene Glycol, Polyglycols, Glycol Ethers, Ethylene Cyanhydrin, and Acrylonitrile, Phenyl Ethyl Alcohol and Related Derivatives of Ethylene Oxide in Germany, (J. D. Brander and R. Max Goepp, Jr.), U.S. Department of Commerce (1946).
8. A. Wurtz, Compt. rend. 50:1195 (1860).
9. S. Perry and H. Hibbert, J. Amer. Chem. Soc. 62:2599 (1940).

10. R. Fordyce, E. L. Lovell, and H. Hibbert, *J. Amer. Chem. Soc.* 61:1905 (1939).
11. P. J. Flory, *J. Amer. Chem. Soc.* 62:1561 (1940).
12. P. A. Levene and A. Walti, *J. Biol. Chem.* 75:325 (1927).
13. I. G. Farbenindustrie, A. G., U.S. Patent Nos. 1,921,378, 1,922,918, and 1,976,678.
14. P. H. Schlosser and K. R. Gray, U.S. Patent No. 2,362,217 assigned to Rayonnier, Inc.
15. H. R. Rife and F. H. Roberts, U.S. Patent Nos. 2,448,664 (1948); 2,520,612 (1950).
16. W. J. Toussaint and H. R. Rife, U.S. Patent No. 2,425,845 (1947).
17. W. H. Millet, *Iron Age Engr.* 25 (8):51 (1948).

BIBLIOGRAPHY

G. O. Curme, Jr., ed., *Glycols*, Reinhold Publishing Company, ACS Monograph, New York, 1952.

A. S. Malinovski, *Epoxides and Their Derivatives*, translated from Russian, 1965 Israel Program for Scientific Translation, Jerusalem, 1965.

4

Polymerization of 1,2-Epoxides

I. INTRODUCTION

The 1,2-epoxides, or vicinal epoxides, include the alkylene oxides, the simplest of which is ethylene oxide. The first reported polymerization of ethylene oxide was by Wurtz in 1863 (1),
who described the reaction of ethylene oxide heated in a sealed
tube with water and, later, with alkali and with zinc chloride
catalysts. Seventy years later, the polymerization by a large
number of catalysts, including alkali and alkaline earth metal
oxides and carbonates, became one of the cornerstones in the
development of the <u>macromolecular</u> <u>hypothesis</u> by Herman
Staudinger (2).

After World War II, very large quantities of three epoxides
were being produced: ethylene oxide, propylene oxide, and
epichlorohydrin. These three epoxides found a wide variety of
uses, including lubricants, surfactants, plasticizers, adhesives,
coatings, and solvents. The first uses involved polymers of
ethylene oxide and propylene oxide with molecular weights in
the range of several hundred to a few thousand as lubricants
and surfactants and epichlorohydrin derivatives in crosslinked
(epoxy) adhesives. During the latter 1950s, polymerization

mechanisms were found that would produce truly high-molecular-weight polymers of vicinal epoxides, as well as ways of controlling the stereoregularity of the polymerization of substituted ethylene oxides and the higher alkylene oxides (3–12). This chapter is concerned with the ring-opening polymerization of 1,2- or vicinal epoxides to produce linear, or other controlled-structure, polyethers.

Three-, four-, five-, or more-membered heterocyclic, aliphatic ring systems with one ring ether oxygen form a series of related, reactive chemical intermediates that can polymerize to form polyethers.

Heterocycle	Relative strain	Oxygen valence angle
1,2-epoxyethane (oxirane)	4.68	61° 24'
1,3-epoxypropane (oxetane)	1.0	94° 30'
1,4-epoxybutane	0.021	111°

While all three ring systems are opened by cationic or acid-initiated mechanisms, only the 1,2-epoxide oxirane can be polymerized by both anionic and cationic initiation. Notice that only 1,2-epoxyethane, ethylene oxide, is derived by epoxidation of an alkene and, thus, is the only alkylene oxide in this series.

Ethylene oxide has the molecular structure shown below.

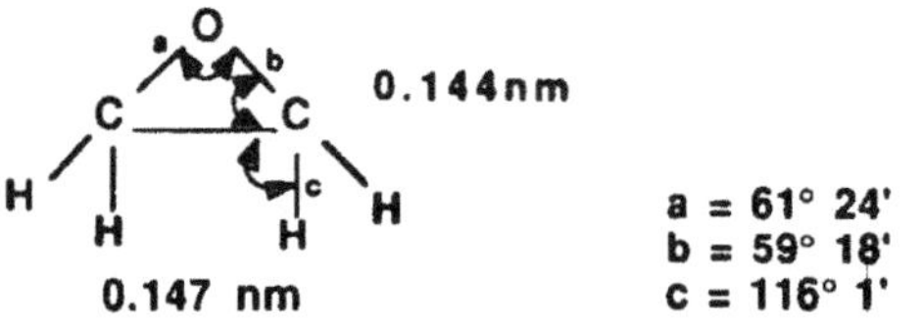

$$a = 61° 24'$$
$$b = 59° 18'$$
$$c = 116° 1'$$

This epoxide ring has an estimated strain energy of 14 kcal/mol (13), and this is the high driving force for ring-opening polymerization to poly(ethylene oxide). The highly exothermic rearrangement of ethylene oxide to acetaldehyde,

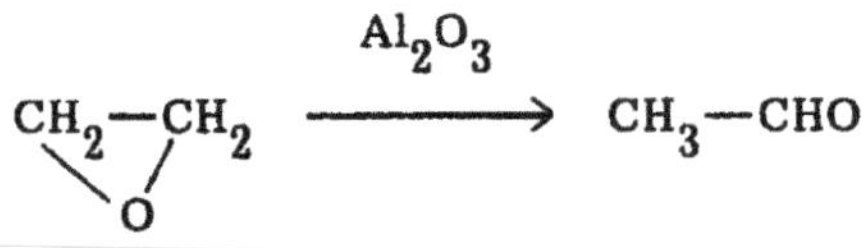

$$CH_2-CH_2 \xrightarrow{Al_2O_3} CH_3-CHO$$

can be catalyzed by alpha-alumina, and this rearrangement was once thought to be an important part of the mechanism by which ethylene oxide polymerized (see Chapter 3). The actual ring-opening polymerization mechanism involves initiation by ionic attack, cationic or anionic, and cleavage of one of the carbon-oxygen bonds rather than rearrangement to acetaldehyde and its polymerization. Free-radical ring opening during hydrogenation of epoxides cleaves the carbon-carbon bond, while in the few documented cases of free-radical-initiated polymerization of epoxides, the reaction involves hydrogen abstraction followed by rearrangements leading to polymers of mixed structure, including polyketones rather than polyethers.

There are three principal ionic ring-opening polymerization reactions of epoxides: acid-initiated, base-initiated, and coordinate-initiated polymerizations. The acid-initiated reaction involves addition of an active hydrogen compound, HY, such as ethanol, to an epoxide ring and is catalyzed by an acid, HX, such as perchloric acid. The reaction sequence involves formation of an oxonium complex, followed by ring opening by an S_N2 cleavage of an oxonium carbon bond.

$$CH_2{-}CHR \atop \diagdown_O\diagup \quad \overset{HClO_4}{\longleftrightarrow} \quad {CH_2{-}CHR \atop \diagdown_O\diagup \atop H^+ \; ClO_4^-} \quad \overset{C_2H_5OH}{\longrightarrow} \quad HO{-}CH_2CRH{-}OC_2H_5$$

monomer   oxonium complex   product (available
                                                         for propagation)

Propagation takes place by addition of n molecules of monomer to the alkoxyethanol product to form $H{-}(OCH_2CHR)_n{-}OC_2H_5$. The rate constant for the reaction relative to ethylene oxide depends on the epoxide or, actually, the substituents on the oxirane ring. For comparisons of polymerization rates for substituted oxiranes, see Table 1.

The ring-opening mechanism for base-initiated polymerization is as follows, where R' is usually an alkyl group such as ethyl.

$$R'O{-}Na \; + \; {CH_2{-}CHR \atop \diagdown_O\diagup} \; \rightarrow \; R'O{-}CH_2CHR{-}O^- \; Na^+$$

base   monomer   initiated species

TABLE 1   Polymerization Rate of Epoxides Relative to
Ethylene Oxide by Simple Acid- or Base-Initiated Reaction (From Refs. 14 and 15.)

| | Initiator | |
Epoxide	$HClO_4$	C_2H_5ONa
Epichlorohydrin	0.065	4.8
Glycidol	0.41	2.0
Ethylene oxide	1.0	1.0
Propylene oxide	54	0.5
Trans-2,3-epoxybutane	119	0.03
Cis-2,3-epoxybutane	238	0.06
Isobutylene oxide	5100	0.06

The ring opening is again of the S_N2 type.   The initiated species will react with additional monomer in the subsequent propagation step to form the polymeric species.   Substituted oxiranes
with electron-withdrawing substituents have higher reactivity
(Table 1).

In the ionic-coordinative polymerization (17, 18) of epoxides,
a metal, M, such as Li, Mg, Zn, Ca, Al, Sn, or Fe, with
ligands, such as OH, OR, Cl, Br, $OSnR_3$, or NH_2, ionically coordinates with the epoxide oxygen, and this is followd by nucleophilic attack of the ligand to open the oxirane ring and form
the initiated species that propagates.

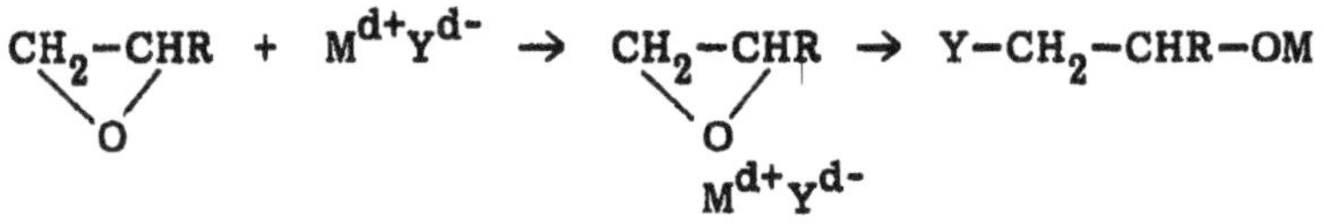

$$CH_2\text{--}CHR \ + \ M^{d+}Y^{d-} \ \rightarrow \ CH_2\text{--}CHR \ \rightarrow \ Y\text{--}CH_2\text{--}CHR\text{--}OM$$

| monomer | initiator | ionic coordina-
tion complex | initiated
species |

II. ACID OR CATIONIC INITIATION OF POLYMERIZATION

The most rapid ring-opening reactions of epoxides are acid catalyzed. The acid may be a Bronsted or protonic acid or a more generalized Lewis acid. Many of the latter are the well-known Friedel-Crafts catalysts, inorganic halides of the form MX_n, such as $AlCl_3$, BF_3, $SnCl_4$, PF_6, and $ZnCl_2$.

The acid- or cationic-initiated polymerization is generally taken to be an S_N2 mechanism. Much early work (19) demonstrated the high yield obtained in such reactions of low-molecular-weight cyclic products, particularly the dimer, dioxane, and substituted dioxanes—i.e., 3,6-dimethyldioxane from propylene oxide—rather than polymeric species.

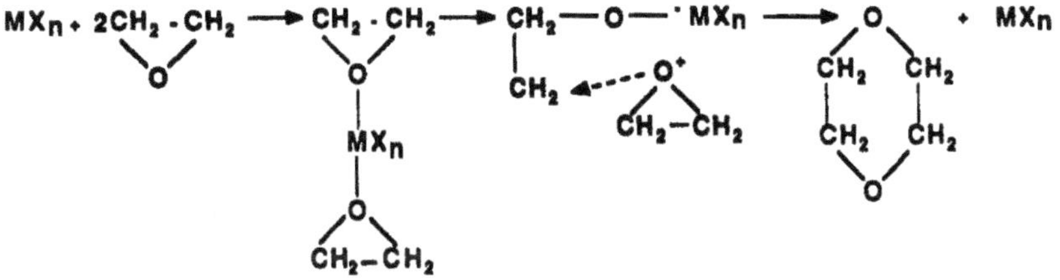

The unifying concept of cationic initiation of polymerization to higher molecular weight products follows from the work of Dainton, Plesch, and Kennedy concerning the role of cocatalysts or formation of the true polymerization initiator (20—22). The experimental result is that the rate of formation of the higher molecular weight, linear polymer and the molecular weight of the polymer produced increase with the concentration of cocatalyst, which may be an alkyl halide or acid chloride or an active hydrogen molecule such as an alcohol or water (23, 24).

Initiator Formation

$$MX_n + ROH \rightarrow MX_n RO^- H^+$$

$$MX_n + RCl \rightarrow MX_n Cl^- R^+$$

Initiation

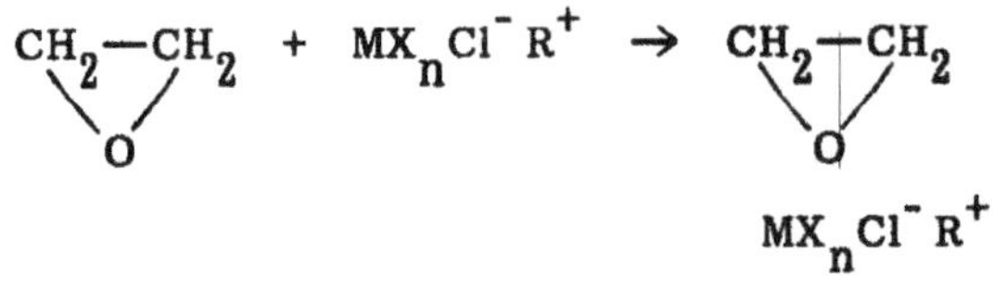

$$CH_2—CH_2 \quad + \quad MX_nCl^-R^+ \quad \rightarrow \quad CH_2—CH_2$$
$$\underset{O}{\diagdown\diagup} \qquad\qquad\qquad\qquad\qquad \underset{O}{\diagdown\diagup}$$
$$MX_nCl^-R^+$$

Propagation

$$RO—CH_2CH_2^+ + \quad MX_nCl^- \quad + \quad x \; CH_2—CH_2 \quad \rightarrow$$
$$\underset{O}{\diagdown\diagup}$$

$$R—(O—CH_2—CH_2)_{(x+1)} \quad + \quad MX_nCl^-$$

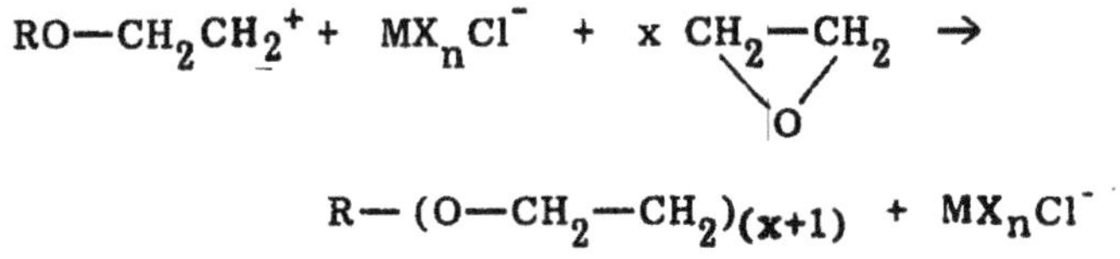

In these reactions, both the rate of polymerization and the average molecular weight of the polymer formed will depend on the
structure of the initiator ion pair and the basicity of the counterion, as well as on predictable temperature and solvent effects.

   Termination of the polymer chain can be proton transfer to
give terminal unsaturation or halogen transfer from the counterion to regenerate the Lewis acid.

$$R—(O—CH_2—CH_2)_{x-1}—O—CH_2—CH_2 \quad + \quad MX_nCl^- \quad \rightarrow$$

$$R—(O—CH_2—CH_2)_{x-1}—O—CH=CH_2 \quad + \quad MX_nHCl$$

   or

$$R—(O—CH_2—CH_2)_xCl \quad + \quad MX_n$$

Polymer molecular weight is determined by the relative rates of
propagation and termination of transfer reactions. Molecular
weight can increase as temperature is raised if the counterion
is sterically large and is a relatively weak base. The molecular
weight distribution is typically broad, and the highest molecular
weight species produced are generally on the order of 10^4 (25)
from acid-initiated polymerizations. In those cases in which the
coinitiator is an active hydrogen molecule, polymerization rate
and molecular weight will depend on the mole ratio of Lewis acid
and coinitiator, since the coinitiator, if in excess, can also
serve as a polymerization chain terminator. In the case of an

alkyl halide, which has been used as the polymerization medium as well as a coinitiator, the alkyl halide can participate in chain transfer as well as in coinitiation.

Imide acids have been used as cationic initiators to study the two potential sites on an oxirane ring where attack by a carboxylic acid can take place (26). In all tests carried out, both normal (C—O) and abnormal (RHC—O) bond cleavage was detected by nuclear magnetic resonance (NMR). As the nucleophilicity of the imide acid increased, the amount of primary hydroxyl produced or of abnormal cleavage that took place was decreased.

An example of cationic ring-opening polymerization of an epoxide is the preparation of polyepichlorohydrin in the presence of diols or water.

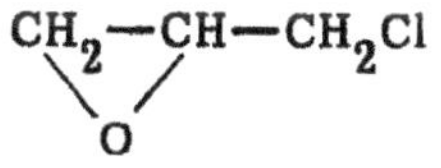

$$CH_2\!\!-\!\!CH\!\!-\!\!CH_2Cl$$
$$\diagdown \diagup$$
$$O$$

epichlorohydrin

Because of the aliphatic chlorine atom, base-catalyzed initiation of epichlorohydrin to relatively high polymer is not practical. Coordinate polymerization is used with the <u>Vandenberg</u> <u>Catalyst</u>, as will be discussed in Section III. Polyepichlorohydrin is used commercially as an oil-resistant rubber.

Polyepichlorohydrin glycols can be prepared in the molecular weight range of 500 to 3000 with relatively narrow molecular weight distributions and weight-to-number average molecular weight ratios of about 1.2 (27, 28). The mechanism is a "living cationic" ring-opening polymerization (22, 29) with molecular weight controlled by the ratio of epichlorohydrin monomer to initiator diol. The initiation of polymerization involves the aklylation of diol—in this example, ethylene glycol—with triethyloxonium hexafluorophosphate.

$$(C_2H_5)_3O^+ \cdot PF_6^- + HOCH_2CH_2OH \;\rightarrow$$

$$C_2H_5\overset{+}{\underset{\mid}{O}}\!\!-\!\!CH_2CH_2OH \;\; PF_6^- + C_2H_5OC_2H_5$$
$$H$$

diethyl ether

This step is followed by reaction of the oxonium salt formed with epichlorohydrin monomer. (The counterion is not shown for purposes of simplicity.)

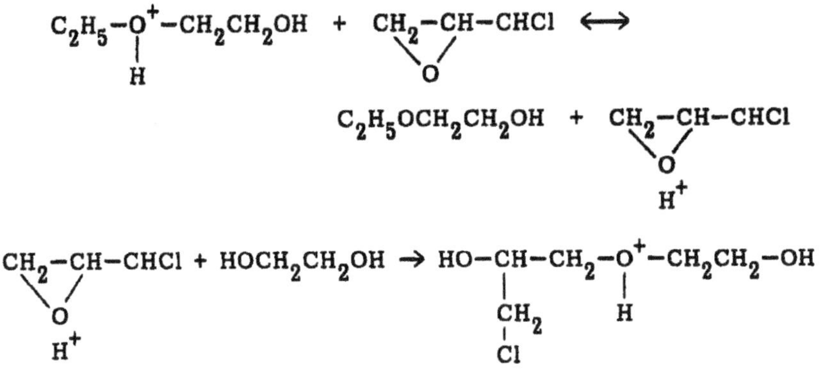

$$C_2H_5-\overset{+}{\underset{H}{O}}-CH_2CH_2OH \ + \ \underset{\underset{O}{\diagdown\diagup}}{CH_2-CH-CHCl} \ \longleftrightarrow$$

$$C_2H_5OCH_2CH_2OH \ + \ \underset{\underset{\overset{O}{\underset{\overset{+}{H}}{}}}{\diagdown\diagup}}{CH_2-CH-CHCl}$$

$$\underset{\underset{O}{\underset{\overset{+}{H}}{\diagdown\diagup}}}{CH_2-CH-CHCl} + HOCH_2CH_2OH \rightarrow HO-\underset{\underset{\overset{CH_2}{|}}{\underset{Cl}{|}}}{CH}-CH_2-\overset{+}{\underset{\underset{H}{|}}{O}}-CH_2CH_2-OH$$

Polymerization proceeds by alternate formation of the oxonium salt of epichlorohydrin, followed by alkylation of the terminal hydroxyl of a growing polymer chain to produce polyepichlorohydrin having the structure

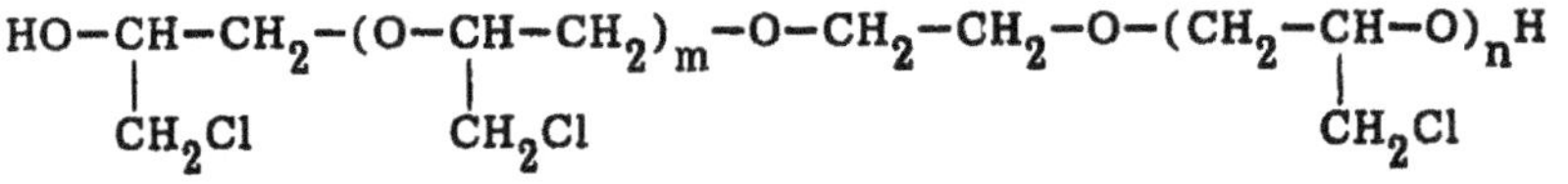

$$HO-\underset{\underset{CH_2Cl}{|}}{CH}-CH_2-(O-\underset{\underset{CH_2Cl}{|}}{CH}-CH_2)_m-O-CH_2-CH_2-O-(CH_2-\underset{\underset{CH_2Cl}{|}}{CH}-O)_nH$$

which is polyepichlorohydrin with a central moiety derived from the glycol initiator. Cationic, photoinitiated polymerization of epichlorohydrin with dicyclopentadienyl-iron complexes has been accomplished with $TiCl_4$, $AlCl_3$, $SnCl_4$, and $FeCl_3$ counterions in the complex (30). With $SbCl_3$ as the counterion, no polymerization occurred.

Characteristics of the polymerization are that all of the ethylene glycol has reacted by the time the polymerization has reached about 20 percent conversion, and the product molecular weight increases linearly with conversion. The terminal hydroxyl groups are principally secondary in nature. Glycols other than ethylene glycol, as well as water, can be used as initiators.

Glycol used	Percent of terminal groups that are secondary OH
Ethylene glycol	93–97
2-butene-1,4-diol	93
water	88–94

Low-molecular-weight (about 500 to 800) polyepichlorohydrin prepared in this manner is relatively free of cyclic oligomers. However, as the molecular weight is increased above this range, a bimodal molecular weight distribution that includes low-molecular-weight cyclic oligomers is formed. Functionality of the polymer formed is in the range of 0.5 to 1.5. If cyclic oligomers are removed by extraction, the linear polymer resulting is essentially difunctional, with the terminal hydroxyl groups secondary in nature. For polymers with average molecular weights of 1000 or more, the cyclic oligomers may constitute about 5 to 20 percent of the product.

III. ANIONIC OR BASIC INITIATION OF POLYMERIZATION

A. Stepwise Anionic Polymerization

Gee et al. studied the kinetics of polymerization of ethylene oxide initiated by sodium methoxide in the presence of a small excess of methanol in dioxane solution (31, 32).

$$CH_3O^- Na^+ + n\ CH_2{-}CH_2 \longrightarrow CH_3{-}O{-}(CH_2{-}CH_2{-}O)_n H$$
$$\underset{O}{\diagdown\diagup}$$

The polymerization rate could be expressed in terms of the disappearance of ethylene oxide monomer

$$-d(EO)/dt\ =\ kC_0(EO)$$

where C_0 is the initial concentration of CH_3ONa. This general expression, which suggests that the reaction involves a closely

bonded ion pair, sodium methoxide, holds for a large number
of alkali/alkoxide-initiated epoxide polymerizations. The activation energy for this reaction was found to be 17.8 kcal/mol.
However, if the same polymerization were carried out in the
presence of an excess of alcohol, the activation energy would be
found to be lower, 14 kcal/mol. The number average degree of
polymerization in the presence of free alcohol could be expressed
as

$$\bar{P}_n = [(EO)_0 - (EO)_t]/[(CH_3OH)_0 + (CH_3ONa)_0]$$

where $\bar{P}_n$ is the number average degree of polymerization, $(EO)_0$
and $(EO)_t$ are the concentrations of ethylene oxide initially and
after reaction time, t, and $(CH_3OH)_0$ and $(CH_3ONa)_0$ are the
initial concentrations of methanol and sodium methoxide.

The reaction rate, in the case of alkali/alkoxide initiation
alone, is quite slow (33), but increases with the concentration
of excess alkanol (34), with a maximum rate at an alkanol/alkox-
ide mole ratio of 1.0. Within this variation in excess alkanol
concentration, the molecular weight distribution of the polyethers produced remains narrow. The rate expression, then,
can be modified:

$$-d(EO)/dt = k_3C_0(EO)(ROH)_0$$

where $(ROH)_0$ is the initial concentration of alkanol, ROH.

These results can be explained by formation of an alkanol/
alkoxide complex as an effective initiator.

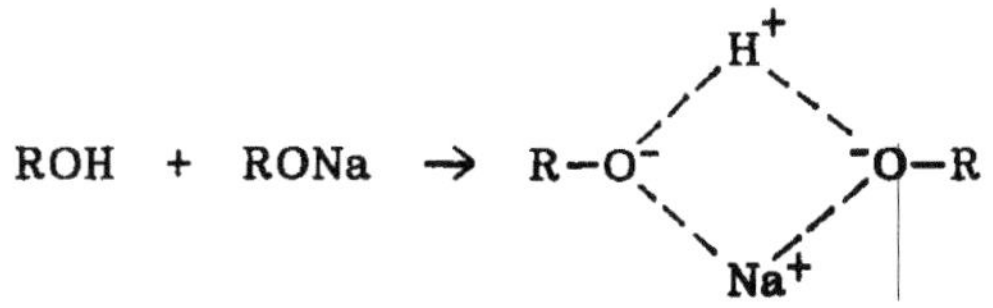

$$ROH + RONa \rightarrow R-O \cdots \overset{\overset{H^+}{/\ \backslash}}{\underset{\underset{Na^+}{\backslash\ /}}{}} \cdots O-R$$

The complex with alkanol is more effective as an initiator than
is the alkali alkoxide alone, because the complexation loosens
the bonding of the alkoxide ion pair. The effective ring-opening reaction, then, must involve a ternary complex of oxirane
and alkanol/alkoxide:

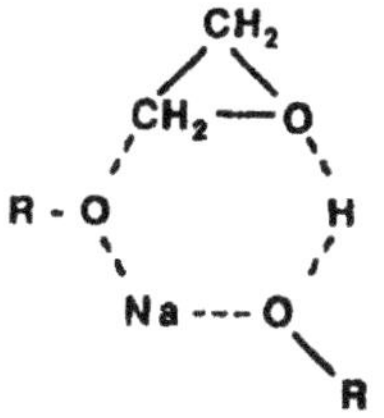

$$
\begin{array}{c}
CH_2 \\
\diagup \;\; \diagdown \\
CH_2 \!\!-\!\! O \\
R\text{-}O \qquad\quad H \\
Na \text{---} O \\
\diagdown \\
R
\end{array}
$$

Using the modified rate equation above, rate constants for addition of ethylene oxide and propylene oxide to a number of alcohols are given in Table 2 (1, 35, 36). From Table 2, several important conclusions can be drawn. As expected, primary hydroxyl groups have a higher reactivity toward oxirane addition than do secondary hydroxyl groups. Second, ethylene oxide addition is considerably faster than that of propylene oxide; third, the rate of addition of the beta-alkoxyethanols is about the same and is not affected by the particular alkoxy substituent.

The terminal groups in the polymerization of ethylene oxide and propylene oxide are principally primary and secondary hydroxyls, respectively (37).

$$ROH + NaOH \longleftrightarrow ROH \cdot NaOH$$

$$[ROH \cdot NaOH] + n\; CH_2 \overset{}{\underset{O}{\diagdown \diagup}} CH_2 \rightarrow RO-(CH_2-CH_2-O)_n H + NaOH$$

$$[ROH \cdot NaOH] + n\; CH_2 \overset{CH_3}{\underset{O}{\diagdown \diagup}} CH \rightarrow RO-(CH_2-\overset{\textstyle |}{\underset{\textstyle CH_3}{CH}}-O)_n H + NaOH$$

With methyl oxirane, propylene oxide, ring opening is by cleavage of the less-substituted carbon-to-oxygen bond. Since the terminal oxyalkanol groups are still active, the polymerizations, as represented, are "living" polymerization reactions. However, there are termination side reactions that become particularly important at elevated reaction temperatures.

With ethylene oxide, the active alkoxide chain end can abstract a hydrogen from the chain, leading to terminal unsaturation and a new chain-initiation site.

TABLE 2 Rate Constants of the Addition of Ethylene Oxide or Propylene Oxide to Various Alcohols

Alcohol	Addition of ethylene oxide		Addition of propylene oxide	
	k_3 at 40°C kg^2/mole·min	E_a, kcal/ mol	k_3 at 55°C kg^2/mol·min	E_a, kcal/ mol
n-propanol	3.89	14.1	0.385	15.4
isopropanol	2.76	14.1	0.124	14.7
n-butanol	3.72	14.1	0.373	14.2
isobutanol	3.43	14.0	0.288	12.1
secbutanol	2.00	14.1		
n-octanol	3.72	14.0		
2-ethylhexanol	1.70			
2-octanol	0.77			
n-propoxyethanol	1.93			
isopropoxyethanol	1.81			
n-butoxyethanol	1.98			
n-octyloxyethanol	2.06			
2-ethylhexyl-oxyethanol	1.93			
n-butoxy-2-ethoxy	2.00			

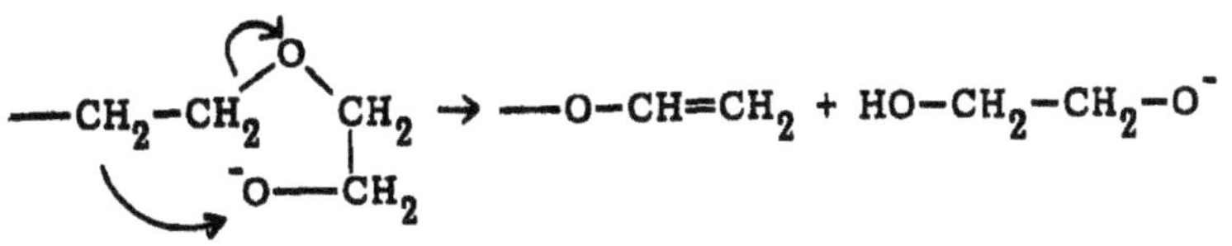

$$-CH_2-CH_2 \overset{O}{\underset{O-CH_2}{\diamond}} CH_2 \rightarrow -O-CH=CH_2 + HO-CH_2-CH_2-O^-$$

With propylene oxide, hydrogen abstraction by the alkoxide oxygen can occur in two ways to give terminal allyl ether or propenyl ether end groups (38, 39).

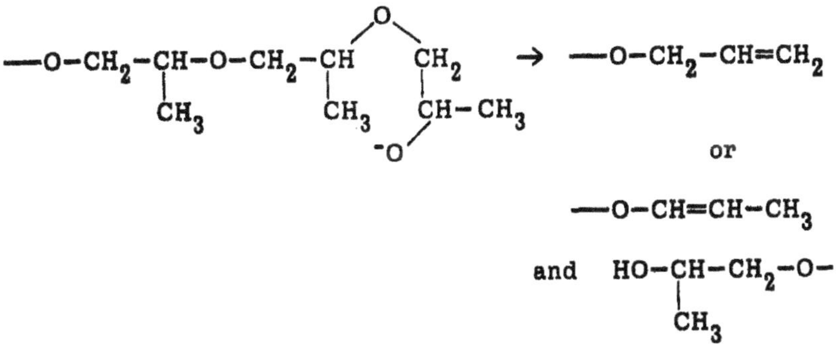

$$-O-CH_2-\underset{CH_3}{CH}-O-CH_2-\underset{CH_3}{CH}\overset{O}{\underset{\underset{-O}{CH-CH_3}}{CH_2}} \rightarrow -O-CH_2-CH=CH_2$$

or

$$-O-CH=CH-CH_3$$

and $$HO-\underset{CH_3}{CH}-CH_2-O-$$

In addition, with propylene oxide, terminal unsaturation can also result from alkoxide ion-induced rearrangement of monomer to produce allyl alcohol, which can then initiate a new polyether chain (40).

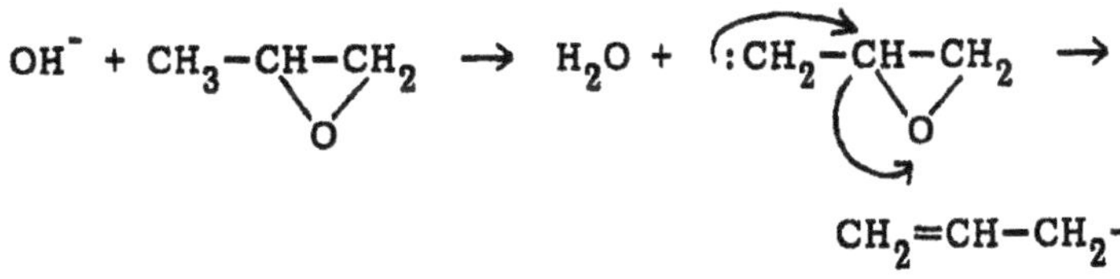

$$OH^- + CH_3-\underset{O}{\overset{}{CH-CH_2}} \rightarrow H_2O + :CH_2-\underset{O}{\overset{}{CH-CH_2}} \rightarrow$$

$$CH_2=CH-CH_2-O^-$$

Under reaction conditions that minimize rearrangement of active species to terminate chain growth and/or generate new hydroxyl-terminated chains, generally lower reaction temperatures (less than 100°C) are needed, and with excess monomer present at all times during the reaction, the number of polymer chains produced is fixed by the number of initiating species present. The stepwise polymerization of epoxides proceeds by the successive addition of monomer to active hydroxyl end groups of polyether polymer chains.

Flory showed that these reaction kinetics lead to polymers with a very narrow molecular weight distribution (41). The reaction criteria for the narrow molecular weight distribution are (1) the growth of each polymer molecule proceeds only by successive addition of epoxide to an active terminal group, (2) all of the active terminal groups are of equal reactivity, and (3) all of the active terminal groups are present initially and their number remains constant throughout the reaction. No chain termination or chain transfer takes place. In the parlance later introduced by Michael Szwarc, these are living polymers (42).

B. Molecular Weight Distribution

The molecular weight distribution of the polymeric compound produced under these reaction conditions and kinetics can be deduced from either statistical or kinetic arguments. The classic statistical argument can be presented as calculating the distribution of missiles hitting each of a fixed number of targets under the conditions that every missile fired hits one target, that the probability of hitting any one target is the same as that of hitting any other target, and that the number of targets is fixed. The resulting distribution of hits on the targets is a Poisson distribution.

If v = the number of missiles fired/number of targets, then the fraction of targets with x hits, N_x/N_0, is

$$N_x/N_0 = v^{x-1}e^{-v}/(x-1)!$$

The kinetic argument that leads to the same molecular weight distribution relates reaction of monomer with time in terms of the concentration of reactive, terminal end groups of polymer chains of varying degrees of polymerization. If N_0 is the number of initiating species at the beginning of the polymerization, (m) is the concentration of monomer at any time t, and N_i is the number of polymer chains with the degree of polymerization i, then a rate equation can be given as follows:

$$-dm/dt = k_0(m)N_0 + k_1(m)N_1 + k_2(m)N_2 + \ldots$$

or, if $k_1 = k_2 = k_3 \ldots$ etc.,

$$-dm/dt = k_0(m)N_0 + k(m) \sum_{i=1}^{i=\infty} N_i = [k_0 + k](m)N_0$$

If, as seen in Table 2, the initiating alcohol is an alkoxy-ethanol, and $k_0 = k$, then the rate expression reduces further to

$$-dm/dt = 2k(m)N_0$$

The ratio of the number of monomer units already polymerized at time t, m, to the number of growing chains, N_0 (also the number of initiator molecules originally present), will be

$$m/N_0 = \int_0^t kN_0(m)dt = v$$

and

$$dv/dt = N_0 k(m)$$

A series of expressions can be written for each species that is formed by reaction of monomer with an active terminal group with one less degree of polymerization and that reacts itself with monomer to form a polymer molecule with one more unit, so that

$$dN_x/dt = -kN_x(m) + kN_{x-1}(m)$$

with the general solution

$$N_x = N_0 v^{x-1} e^{-v}/(x-1)!$$

which represents a Poisson distribution.

A Poisson distribution of molecular weights is very narrow. The consequences of a Poisson distribution can be seen in Figure 1, in which examples of a Poisson distribution of molecular weight are compared, at different degrees of polymerization, with a most probable distribution, a Gaussian distribution, in which the ratio of the weight to number average molecular

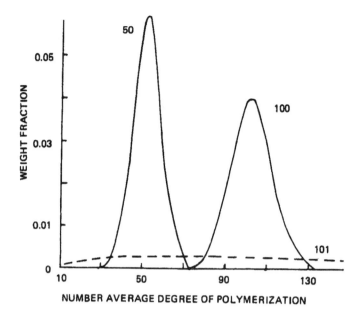

FIG. 1 Comparison of molecular weight distributions. Poisson
distributions (———) and Gaussian or most probable distribution (-----) for indicated numbers of monomer units reacted per
initiator molecule. (From Ref. 41.)

weights approaches a value of 2. A most probable or Gaussian
distribution of molecular weight arises, for example, from a condensation polymerization of a dibasic acid and a diol or from addition polymerizations that propagate without chain transfer (43,
44). For many applications, such as surfactants and lubricants,
the narrow molecular weight distribution of the poly(alkylene
oxides) is an important factor in improving product performance.

While a Poisson distribution of molecular weights results
when the rates of additon of epoxide to terminal hydroxyls are
all equal, it is clear from Table 2 that this case generally occurs only when the initiating alcohol is an alkoxyethanol.

$$R—O—CH_2CH_2OH$$

In other cases, the first alkoxylation step is generally slower
than the subsequent polymerization steps. This circumstance

TABLE 3   Distribution Constants in Polym-
erization of Ethylene Oxide Initiated by
Various Alcohols in the Presence of 1.0%
NaOH at 40°C.   Distribution Constant,
$C = k/k_0$

Starting alcohol	C
ethanol	1.93
n-propanol	3.01
isopropanol	11.2
n-butanol	3.40
isobutanol	4.11
secbutanol	25.9
t-butanol	ca. 60
n-octanol	2.78
n-butoxyethanol	1.1
n-dodecyloxyethanol	1.2

leads to some broadening of the molecular weight distribution.
An expression for this modified molecular weight distribution
was formulated by Weibull and Nycander (45), and experimental
confirmation was obtained by Nishii et al. (46) in terms of distribution constant, C, the ratio of the relative reactivities of
the oxyalkyl alcohol product and the starting alcohol (Table 3).
The order of the distribution constant follows the expected order of starting alcohol reactivity, primary < secondary < tertiary.

IV.   COORDINATE-INITIATED POLYMERIZATION

While a number of the early systems for the polymerization of
ethylene oxide and propylene oxide might now be categorized as
coordinate initiated (Table 1), the first of the more recently
discussed systems that opened the way for a better understanding of these systems was found by Pruitt and Baggett in 1952

(7). This polymerization system, a complex of ferric chloride and propylene oxide, produced high-molecular-weight polymer rapidly. Further, a significant part of the poly(propylene oxide) formed, about 30 percent of the product, was found to be crystalline (i.e., acetone insoluble), with a melting point of about 75°C.

Subsequently, a large number of initiator systems have been found. Some of these are listed in Table 4, in which the initiators are placed in one of two categories following the work of Furukawa and Saegusa (47, 54). In the first category, initiation involves formation of a metal-oxygen coordination bond. In the second category, alkaline earth compounds present an anionic surface that apparently can coordinate with oxirane monomer, followed by nucleophilic attack of an alkoxide ligand and ring opening of the oxirane (60—62). Both groups are effective in polymerizing ethylene oxide to high polymer. In general, the alkaline earth compounds lead to a more rapid ethylene oxide polymerization than do the other compounds. The first category, which involves a metal-oxygen coordination, is generally more effective than the second category in initiating propylene and butylene oxide polymerization and in effecting stereoregular control of the polymer formed from substituted oxiranes.

The mechanism of the ferric chloride-propylene oxide complex-initiated polymerization has been considered to be a coordinate anionic polymerization in which monomer, propylene oxide, first coordinates with iron, followed by nucleophilic attack of the alkoxide group on one of the oxirane carbons.

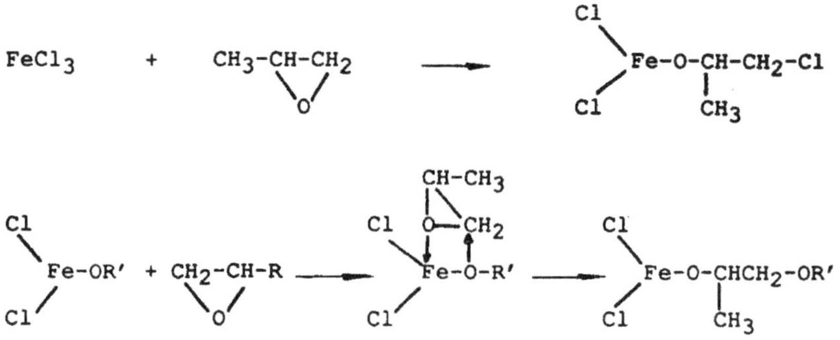

$$FeCl_3 \quad + \quad CH_3-CH-CH_2 \quad \longrightarrow \quad \underset{Cl}{\overset{Cl}{>}}Fe-O-CH-CH_2-Cl$$

$$\underset{Cl}{\overset{Cl}{>}}Fe-OR' + CH_2-CH-R \longrightarrow \underset{Cl}{\overset{Cl}{>}}Fe-O-R' \longrightarrow \underset{Cl}{\overset{Cl}{>}}Fe-O-CHCH_2-OR'$$

The active initiator is first formed by reaction of propylene oxide with ferric chloride at 50—60°C, followed by polymerization of propylene oxide at 80°C. The polymer obtained was of very

TABLE 4 Polymerization Initiators for Alkylene Oxides

Category I Initiators	Example	Reference
Group II and III metal alkoxides	$Al(OR)_3$, $Zn(OR)_2$	(48, 49)
Complex metal alkoxides	$Ca_{0.5}[Al(OR)_4]$	(50)
Metal alkoxide/salts	$Al(OR)_3 \cdot ZnCl_2$	(51, 52)
Metal alkoxide/water	$Al(OR)_3 \cdot H_2O$	(53)
Ferric chloride/epoxide	$FeCl_3$/propylene oxide	(7)
Organometallics	MgR_2, $RMgX$	(55, 56)
Organometallic/metal halide	$ZnR_2 \cdot CaF_2$	(57)
Organometallic chelate	$MgR-2 \cdot$ acetylacetone	(58)
Acidic oxide/organometallic	$Al_2O_3 \cdot ZnR_2$	(59)
Metallic oxide/zinc alkyl	$ZnEt_2 \cdot MgO$	(57)
Category II Initiators		
Alkaline earth oxides	CaO	(63)
Alkaline earth carbonates	$SrCO_3$	(64)
Alkaline earth alkoxides	$Sr(OR)_2$	(65)
Alkaline earth amide-alkoxides	$Ca(OR)NH_2$	(66)
Alkaline earth amides	$Ca(NH_2)_2$	(67)
Alkaline earth hexammoniate	$Ca(NH_3)_6$	(68)
Alkaline earth chelates	Ba acetoacetate	(69)

high molecular weight, with an acetone-insoluble fraction, about
30 percent of the product, having a crystalline melting point of
74—75°C. The active species in the polymerization was extensively studied (8, 67, 68), and it was concluded that the mechanism of polymerization was of the alkoxide type as shown above.

Propylene oxide has an asymmetric carbon atom. The normal commercial epoxide is a racemic mixture of the d- and l-iso-
mers. Osgan and Price did extensive work with both the
l-propylene oxide and the d,l-propylene oxide in both potassium
hydroxide and ferric chloride/propylene oxide-initiated polymerizations. Their results are summarized in Table 5 (48). C. C.
Price and coworkers first demonstrated that polymerization of
pure l-propylene oxide with an anhydrous potassium hydroxide
(solid KOH) initiator led to a crystalline, rather than the usual
amorphous, liquid, polymer. After extensive study by a number of researchers (69), this polymerization was shown to proceed by a stepwise anionic mechanism. The uses found for
polymers of propylene oxide largely have been those requiring
the amorphous polymer in elastomeric applications. Stereospecificity, however, has proved to be a key tool in understanding the polymerization mechanisms.

It was later shown that a small amount of water added to
the FeCl$_3$/propylene oxide initiator system increased both the
polymerization rate and the amount of crystalline product formed
(67, 70) (Figure 2). This effect of an added reactive species
such as water, an alcohol, or acetylacetone to one of the stereospecific initiator structures having a strong influence on polymerization rate and stereospecificity has been found in a number
of the important oxirane polymerization systems.

A. Configurations of Poly(propylene oxide)

Most ring-opening polymerizations present no difficulties in ascertaining the configuration of the polymer produced. The configuration is evident from the monomer and polymerization mechanism. However, propylene oxide is an exception to this usual
simplicity because both monomer and polymer contain true asymmetric centers, and the configuration of polymer presents a
complex challenge in stereochemistry. In the next section
(Section IV-B), it will be found that the solution to the configurations of the polymers of the alkylene oxides leads to an understanding of the polymerization mechanisms, a development
that is one of the most elegant chapters in polymer synthesis.
The aluminum (71), zinc (72), and calcium (73) catalysts can

TABLE 5 Comparison of Polymers Prepared by Potassium Hydroxide and by Ferric Chloride/Propylene Oxide Complex Initiated Polymerizations of 1-Propylene Oxide and d,l-Propylene Oxide

Initiator	KOH		FeCl$_3$/PO			
Epoxide	l-PO	d,l-PO	1-PO		d,l-PO	
Polymer			Unfractionated	Crystalline	Amorphous	Crystalline
[n]D in CHCl$_3$*	+25	—	+17	+25	+3	—
Tm†, °C	56	Liquid	74	74	—	74
Intrinsic viscosity	0.1	0.1	3.1	3.1	1.4	3.1

*Optical rotation in chloroform.
†Melting point.

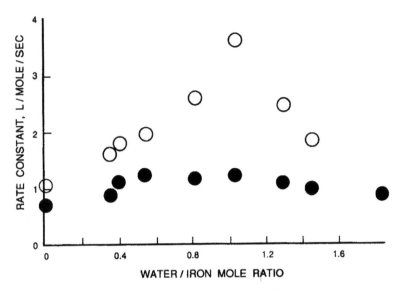

FIG. 2 Effect of water on the polymerization of ethylene oxide
and propylene oxide initiated by FeCl₃ (From Ref. 70.) ○ =
propylene oxide, and ● = ethylene oxide

be considered <u>inorganic</u> <u>enzyme</u> models in their stereoselectivity
and control of polymer structure. This promising anology,
pioneered in polymer chemistry of the alkylene oxides, now
ranks among the most active areas in organic synthesis (73—75).
     The monomer, propylene oxide

$$CH_2\underset{\diagdown O \diagup}{-}C^*H-CH_3$$

has an asymmetric carbon atom, C^*. When the ring is opened,
this carbon remains asymmetric in the polyether chain. Having
asymmetric centers, the polymer chain has a sense of direction,
so that along the chain, a sequence of structural units "ddl" is

not equivalent to an "lld" sequence. Further, both head-to-head

$$-CH_2-CH(CH_3)-O-CH(CH_3)-CH_2-O-$$

and head-to-tail

$$-CH_2-CH(CH_3)-O-CH_2-CH(CH_3)-O-$$

sequences can occur during polymerization. The latter sequences are formed by the expected carbon-oxygen bond cleavage between the methylene group and the oxirane oxygen, and the former by abnormal cleavage between the heavily substituted or asymmetric carbon atom and the oxirane oxygen.

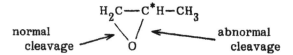

Usually, carbon-13 NMR is used to determine both the asymmetry and head-to-head isomerism, but both situations seldom arise in the same polymer system. The head-to-head isomerism can affect chemical shifts to at least the same degree as does the stereochemistry.

In poly(propylene oxide), both crystalline and amorphous polymer can occur. Amorphous polymer, or amorphous segments in the polymer chain, can arise either from atactic sequences along the chain (stereochemistry) or from head-to-head isomerism in an otherwise tactic sequence. Both Vandenberg (76) and Price (77) found up to 30 percent head-to-head structures in some poly(propylene oxide)s. Polymer from mixed abnormal and normal ring opening of propylene oxide is most likely to occur when there is considerable cationic character to a coordination polymerization initiator.

Considering only the normal ring opening, cleavage of the less-substituted carbon-oxygen bond, there are four triad sequences possible. If the plane projections of the molecules are viewed, one finds the following, in which R represents the methyl group of propylene oxide and the alkyl group of other alkylene oxides.

isotactic:
$$\begin{array}{ccccccccc} & H & R & & H & H & & H & R \\ & | & | & & | & | & & | & | \\ -C & - & C & - O - & C & - & C & - O - & C & - & C & - O - \\ & | & | & & | & | & & | & | \\ & H & H & & H & R & & H & H \end{array}$$

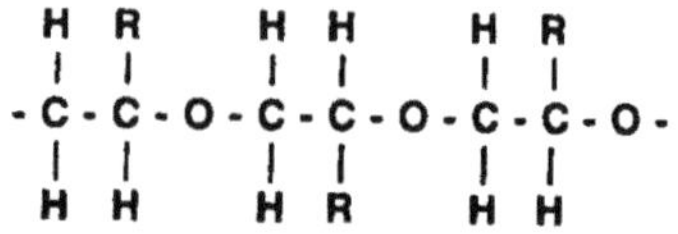

heterotactic 1:
$$\begin{array}{ccccccccc} & H & H & & H & H & & H & R \\ & | & | & & | & | & & | & | \\ -C & - & C & - O - & C & - & C & - O - & C & - & C & - O - \\ & | & | & & | & | & & | & | \\ & H & R & & H & R & & H & H \end{array}$$

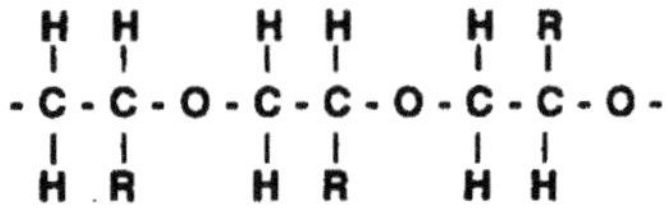

heterotactic 2:
$$\begin{array}{ccccccccc} & H & R & & H & H & & H & H \\ & | & | & & | & | & & | & | \\ -C & - & C & - O - & C & - & C & - O - & C & - & C & - O - \\ & | & | & & | & | & & | & | \\ & H & H & & H & R & & H & R \end{array}$$

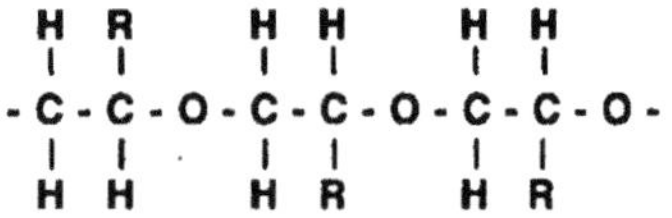

syndiotactic:
$$\begin{array}{ccccccccc} & H & R & & H & R & & H & R \\ & | & | & & | & | & & | & | \\ -C & - & C & - O - & C & - & C & - O - & C & - & C & - O - \\ & | & | & & | & | & & | & | \\ & H & H & & H & H & & H & H \end{array}$$

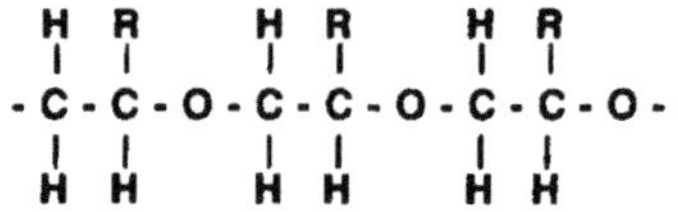

Because the polyether chain has three atoms per structural unit, the R group (methyl, in the case of propylene oxide) appears alternately on opposite sides of the zigzag plane in the isotactic form and on the same side in the syndiotactic form. These configurations also can be represented in the following way.

isotactic:

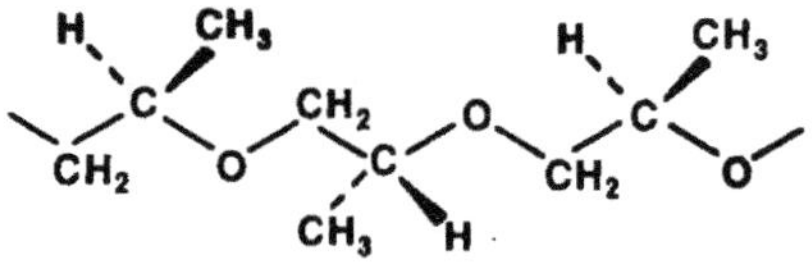

syndiotactic:

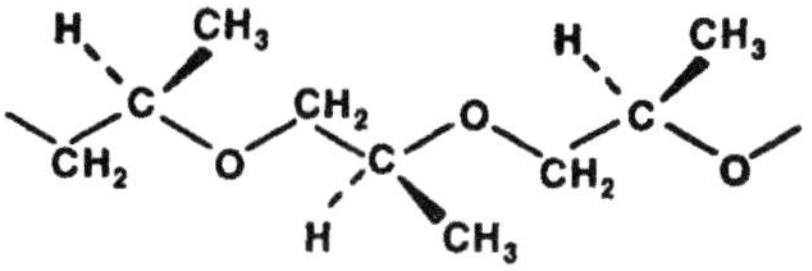

Anticipating Section IV.D, tacticity can come about because the initiator is enantiomorphic selective or because incoming monomer is controlled by the configuration of the terminal or

penultimate structural unit of polymer. Stereospecific olefin
polymerizations are generally of the latter type, while both
types probably have been observed with the alkylene oxides.
In the case of propylene oxide, where the crystalline polymer
arises from enantiomorphic selectivity of the site for monomer
coordination, the isotactic configuration is the more likely to occur. Syndiotactic polymer would result from perfectly alternating monomer selectivity during propagation. Stereoblock polymer, in which more extended sequences of one antipode monomer
would be followed by an extended sequence of the other, would
lead to crystalline, but optically inactive, polymer, from a racemic mixture of monomers (78).

Control of enantiomorphic selectivity in polymerization of
the substituted oxiranes can lead to controlled-structure polymers, the properties of which will range from crystalline ther-
moplastics to amorphous elastomer precursors such as are used
as soft segments in polyurethanes. Crystallizable sequence
distributions in highly controlled-structure polymers can lead to
thermoplastic elastomers and/or to elastomers that will stress-
crystallize; that is, crystallize on stretching as does natural
rubber (79).

In the next section, polymerization initiators that are enantiomorphic selective in producing poly(alkylene oxides) with
highly controlled structure will be discussed. These will include the Vandenberg, Tsuruta, and Bailey catalysts, which
display controlled <u>molecular</u> <u>recognition</u> <u>synthesis</u>. Generally,
the polymerization mechanism involves opening of the oxirane
ring by rearward attack on the less substituted carbon and inversion of configuration of the ring-opening carbon atom. It
is highly probable that the transition state of the monomer en-
tering the growing polymer chain involves a three-centered,
planar complex to permit this controlled configuration inversion.
It is interesting to draw an analogy between these enantiomor-
phic-selective polymerizations and the 1948 suggestion of Linus
Pauling, differentiating the bioactivity of antibodies and enzymes, that antibodies bind molecules in their ground state
while enzymes bind the transition state, thus stabilizing the
configuration relative to substrate and product. In the en-
antiomorphic-selective polymerization of the alkylene oxides, the
polymerization mechanism must involve complexing of a particular
antipode monomer in a linear or planar transition state, followed
by ring-opening addition.

B. Group II and Group III Organometallic Initiators

Early work identified Group II organometallic compounds such as diethylzinc and ethylmagnesium bromide as coordinate initiators of oxirane polymerization. Under carefully controlled conditions with anhydrous, high-purity monomer, the initiator activity was usually found to be weak. Bailey (80) found that with propylene oxide, distilled and maintained anhydrous with molecular sieves, the rate of polymerization of propylene oxide with dibutylzinc actually decreased as the concentration of dibutylzinc increased. This observation strongly suggested that a trace cocatalyst was needed to effect polymerization.

Furukawa and coworkers demonstrated that the effectiveness of diethylzinc increased with controlled addition of water, methanol, or oxygen (81, 82) (Tables 6 and 7). These polymerization systems with propylene oxide produced a substantial fraction of crystalline polymer (16 percent in the case of water/diethylzinc). In some cases, poly(propylene oxide) was prepared with molecular weights estimated to be tens of millions (84). Efforts to understand these polymerization systems centered on identifying the coordination sites available as initiation sites on different metal alkoxides and on differentiating cationic initiation systems from coordinate initiation systems. In some cases, very similar systems led to quite different polymerization reactions and sometimes very different polymer products.

Even though very high-molecular-weight and partially crystalline polymer was formed with propylene oxide, the molecular weight of the polymer increased with the degree of reaction, and polymerization could be reinitiated by addition of more mon-

TABLE 6   Polymerization of Propylene Oxide by Dibutylzinc/Water (From Ref. 81.)

Mole ratio, $H_2O/ZnEt_2$	Polymer yield, room temperature for 24 hours	Intrinsic viscosity, in benzene at 30°C
0	0	—
0.50	10.7%	2.3
1.0	78.3%	5.4
2.0	0	—

TABLE 7   Activity of Diethylzinc/Coinitiator Systems with Propylene Oxide at 70°C for 24 Hours

Coinitiator	Optimum mole ratio, coinitiator/water	Polymer yield, %	Int. visc.*	Reference
Water	1.0	85	5.4	(81)
Methanol	2.0	54	1.4	(81)
Ethanol	2.0	8	1.2	(83)
Glycerol	2/3	81	4.8	(83)
Ethylene glycol	1.0	51	3.5	(83)
Acetone	1.0	86	3.2	(83)

*Intrinsic viscosity in benzene at 25°C

omer, all of which were evidence of a living, stepwise anionic mechanism. In some cases, the molecular weight distribution of the polymer was broad. Broad molecular weight distributions are not expected from polymers prepared via an anionic, living growth mechanism. Trialkyl-aluminum/water as an initiator apparently could initiate a coordinate cationic ring-opening polymerization, while both dialkylzinc/water and dialkylmagnesium/water initiators led to coordinate initiation. With the diethylzinc/water system, the active species is postulated to be

$$[(Et-Zn-OH)_m(ZnO)_n]$$

A mechanism for the initiation of polymerization of propylene oxide from such a species was proposed involving a four-member ring intermediate:

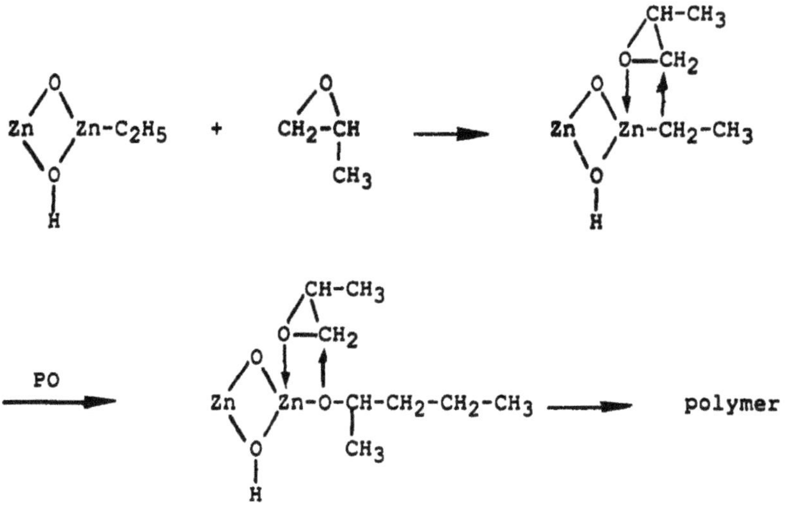

Zinc alkoxides, $Zn((OR)_2$, were found to be effective initiators of polymerization under certain circumstances. Formed in situ as the methoxide or ethoxide, these zinc compounds were effective when amorphous. The crystalline forms were much less effective (85). Zinc alkoxides formed in situ could be considered monomeric, dimeric, or more highly associated.

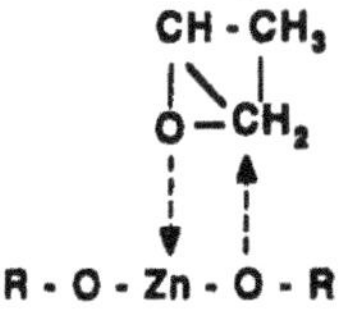

Monomeric Zinc Alkoxide Coordinated
with Propylene Oxide

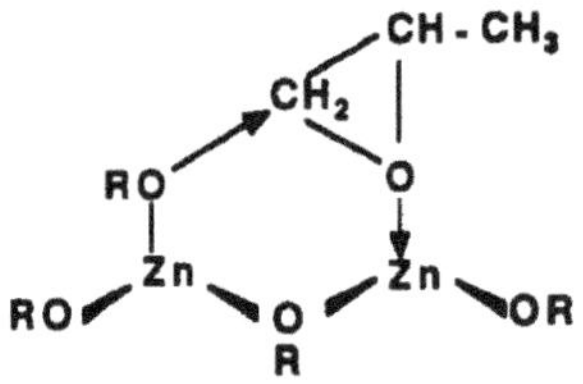

Zinc Alkoxide Dimer Coordinated
with Propylene Oxide

Both the monomeric and dimeric species could be considered to form four-centered intermediates as indicated.

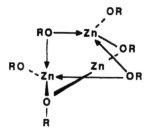

Zinc Alkoxide Trimer

However, the trimeric form, although amorphous with the zinc atoms coordinated tetrahedrally, was a relatively low-activity initiator apparently having no readily available coordination sites for monomer. The trimeric form and higher associated forms of alkoxide tend to form with time as the alkoxide, formed in situ, associates or flocculates on standing, and it was this observation that was believed to explain the reduction in activity observed with some initiators on standing for some time before use.

Acetone was used as a coinitiator with diethylzinc. The active species appeared to be

$$CH_2=C(CH_3)-O-Zn-Et.$$

Oxygen as a coinitiator for diethylzinc led to a mixed alkoxide (86):

$$Et-O-Zn-Et \quad and \quad (EtO)_2Zn$$

A zinc alkoxide coordination initiator that has an amorphous, heterogeneous composition can be represented as

$$[Zn(OR)_2]_n(RO-Zn-OR)$$

in order to focus on the activity of one zinc alkoxide moiety.

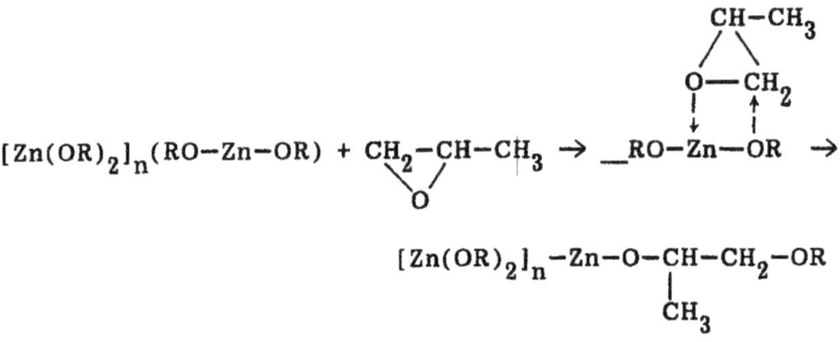

$$[Zn(OR)_2]_n(RO-Zn-OR) + \underset{\underset{O}{\diagdown\diagup}}{CH_2-CH-CH_3} \rightarrow \underline{\quad}RO-Zn-OR \rightarrow$$

$$[Zn(OR)_2]_n-Zn-O-\underset{\underset{CH_3}{|}}{CH}-CH_2-OR$$

The reaction can produce a number of results:

The resulting polymer can be amorphous, with no indication of
   stereospecificity or optical isomer selectivity.
The resulting polymer can be crystalline, suggesting stereospecificity but no optical selectivity.
The resulting polymer can be both crystalline and optically active, suggesting that the polymerization was both stereospecific and enantiomorphic selective.

Further possibilities are that the amorphous polymer can result
if the initiation sites are nonselective and/or the initiation sites
are permitted abnormal oxirane ring fission.

   The mechanism, following Tsuruta given above, indicates
that normal ring cleavage of the less-substituted carbon-oxygen
bond of the oxirane ring will take place. However, abnormal
cleavage can occur, and it gives rise to head-to-head and tail-
to-tail polymer chain units (76, 77).

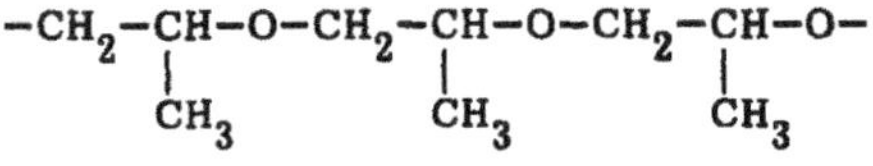

$$-CH_2-\underset{\underset{CH_3}{|}}{CH}-O-CH_2-\underset{\underset{CH_3}{|}}{CH}-O-CH_2-\underset{\underset{CH_3}{|}}{CH}-O- \qquad -CH_2-\underset{\underset{CH_3}{|}}{CH}-O-\underset{\underset{CH_3}{|}}{CH}-CH_2-O-\underset{\underset{CH_3}{|}}{CH}-CH_2-O-$$

polymer from normal   polymer from mixed abnormal-
oxirane ring opening   normal oxirane ring opening

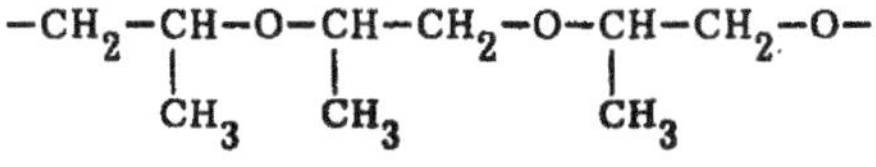

   Potassium hydroxide initiator produces crystalline poly(propylene oxide) from l-propylene oxide, but amorphous polymer
from the racemic, d,l-monomer. Price and Spector (77) calculated the amount of abnormal ring cleavage that occurred with
several initiator systems from the difference in optical rotation
of polymer obtained with pure optical isomer monomer and these

initiators and polymer that was obtained by polymerization of
the pure isomer monomer with stepwise anionic addition. This
calculated result was confirmed by chromatographic analysis of
the glycols obtained by reduction of the ozone oxidation products of the polymers with lithium aluminum hydride. The results indicated that some initiator systems, such as diethylzinc/
water, were essentially stereospecific with nil abnormal ring
cleavage. Further, a number of coordination initiators, including $FeCl_3$/PO and $ZnEt_2$/H_2O, produced crystalline polymer both
from pure optical isomer monomer and from d,l-monomer.

It is possible, then, to produce crystalline, stereospecific
poly(propylene oxide) from optically pure isomer and racemic
monomer with a variety of initiators (77) and to determine the
product distribution of crystalline and amorphous fractions as
well as the optical activity of the products. Further, in a penetrating series of experiments, Furukawa, Tsuruta, and coworkers used optically active coinitiators, including menthol and
borneol, and optically active alkylzinc, di-2-methylbutylzinc/
H_2O, to prepare crystalline polymer with optical activity from
d,l-monomer, observing enrichment of unreacted monomer with
the unreacted enantiomer monomeric species (79, 89—92).

The conclusion was that with some initiator systems, abnormal ring cleavage can occur in the polymerization of propylene
oxide. However, with some initiator systems, notably dialkyl-
zinc/water or some alcohols, the initiation sites are strongly
stereoselective, producing stereospecific polymer of narrow molecular weight distribution. The selectivity appears to occur in
the monomer coordination step at the active initiator site.

Considerable confusion arises in the literature concerning
the stereochemistry of the polymerization of epoxides because
of an early error in the assignment of the configuration of l-
propylene oxide. The analysis of the configuration of the
epoxides and their polymers was undertaken before carbon-13
NMR was fully available for such analysis. As a result, heroic
means, principally due to Vandenberg, were required before
the stereomechanisms could be understood. It was, indeed, un-
clear at the time whether, in the preparation of high-molecular-
weight polymer by various initiators, the terminal-group hydroxyls were principally primary or secondary. In the following
discussions, the work of many chemists is translated into the
stereochemistry as it was finally understood. The reader is
warned, however, that in returning to original references, it is
necessary to ascertain which definitions of configuration are
used. [(+)PO belongs to the d-series and has an "R" configuration.

Vandenberg discovered a class of initiators for the polymerization of epoxides that was a further extension of the trialkylaluminum/water systems (93). This work was an outgrowth of his work with coordination polymerization, or insertion polymerization, of olefins. The principal initiator of this group was the chelate or Vandenberg Catalyst, which combined water-complexed trialkylaluminum and acetylacetonate. This initiator proved particularly effective in the polymerization of epichlorohydrin and in providing a new commercial class of polyether elastomers that BF Goodrich commercialized, under license, and sold under the trademark HYDRIN. The preparation of the Vandenberg Catalyst can be described by the following equations.

$$2\ R_3Al + H_2O \longrightarrow R_2Al\text{-}O\text{-}AlR_2 \quad + \quad 2RH$$

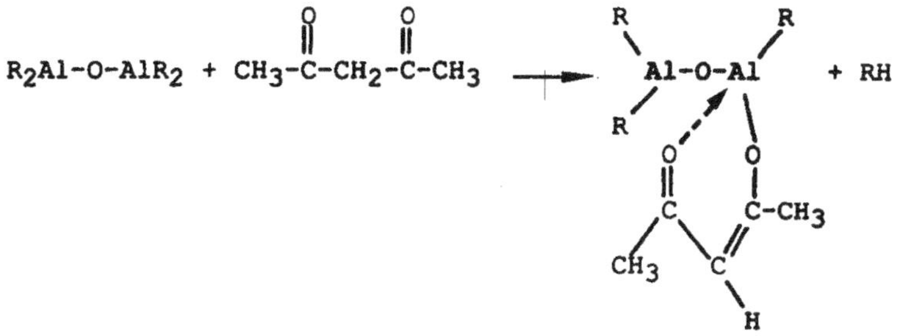

$$R_2Al\text{-}O\text{-}AlR_2 + CH_3\text{-}C\text{-}CH_2\text{-}C\text{-}CH_3 \longrightarrow$$

With this initiator, Vandenberg was also able to find a route to sulfur-vulcanizable polyether elastomers through the copolymerization of propylene oxide with allyl glycidyl ether (71, 94), which Hercules sold commercially under the trademark PAREL. Both of these useful elastomers were amorphous—that is, atactic—polymer structures.

The range of polymer properties that can be achieved depending on the stereochemistry of the poly(alkylene oxide) is in part exemplified by poly(propylene oxide), but can be extended to the entire spectrum of polymers of alkylene oxides (Table 8).

Vandenberg, following a particularly penetrating line of research, used the polymerization of cis- and trans-2,3-epoxybutane to distinguish the polymerization mechanisms of oxirane coordination polymerization and then to generalize this basic mechanism to monosubstituted oxiranes (95). The mechanism

TABLE 8   Range of Polymer Structure/Properties Through
Stereo-Control of Epoxide Polymerization

Monomer	Initiator	X-ray crystallinity	Tm, °C
Propylene oxide	$ZnEt_2/H_2O$	High	74
	Al (i-Bu)$_3$	Low	—
Epichlorohydrin	Al (i-Bu)$_3$/VCl$_3$	High	120
	Al (i-Bu)$_3$/H$_2O$	Low	—
Cis-2,3-epoxy-butane	Al (i-Bu)$_3$	Amorphous	—
	Vandenberg Catalyst	High	162
Trans-2,3-epoxy-butane	Al (i-Bu)$_3$	High	100
	Vandenberg Catalyst	Very low yield	100
Cyclohexene oxide	Vandenberg Catalyst	Crystalline fraction	76
Phenyl glycidyl ether	AlEt$_3$/H$_2O$	Crystalline fraction	203

can account for the very different products obtained from the
polymerization of one monomer.

   2,3-Epoxybutane can have either a cis or trans configuration.   Further, both carbon atoms of the oxirane ring are asymmetric.   Depending on polymerization initiator, monomer, and
reaction conditions, homopolymerization can lead to amorphous or
either of two crystalline polymers, as described in Table 9 (95,
96).   The polymer configurations were established by cleaving
the polyether chains and identifying the stereoconfigurations of
the glycols obtained.

   The conclusion was that all polymerizations occurred with
inversion of configuration of the ring-opening carbon atom.
This generalization implies that the polymerization step cannot

TABLE 9 Polymerization of 2,3-Epoxybutane

2,3-Epoxy-butane	Initiator	Polymerization temperature, °C	Polymer
Cis-monomer	R_3Al/H_2O	−78, Very fast	Amorphous, disyndiotactic
Cis-monomer	Vandenberg Catalyst	65 Slow reaction	Crystalline, mp = 162°C dl-racemic, diisotactic
Trans-monomer	R_3Al/H_2O	−78 Very fast	Crystalline, mp = 100°C, meso-diiso-tactic
Trans-monomer	Vandenberg Catalyst	65 Small yield	Crystalline, mp = 100°C, meso-diiso-tactic

proceed through a four-centered-ring intermediate with a single metal atom, since such an intermediate would not permit the linear transition state necessary for configuration inversion. The mechanism must accommodate the known terminal groups, secondary hydroxyls, to be formed in the polymerization and also the principal cleavage of the oxygen/less-substituted carbon bond, in the case of the monoalkyl-oxiranes. Vandenberg proposed that in the case of coordination polymerization, the active initiating site for both mono- and disubstituted oxirane polymerization had to involve two metal atoms. The mechanism that Vandengerg proposed was shown to be valid also for ethylene oxide by using NMR to observe the polymerization products of cis- and trans-dideuterooxyethylene, which gave results entirely analogous to the 2,3-epoxybutanes (77).

With propylene oxide and aluminum-based initiators, such as the Vandenberg Catalyst, the mechanism proposed can be represented as an alkoxide and an oxirane monomer associated with neighboring four-coordinated aluminum atoms with the approach to the site of a second propylene oxide monomer molecule.

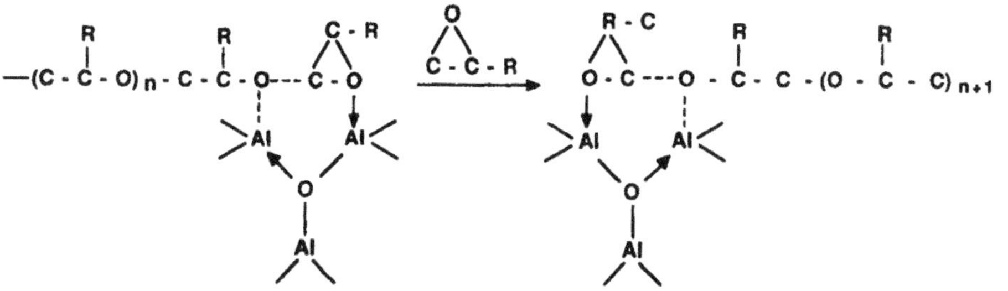

$$-(C-C-O)_n-C-C-O---C-O \xrightarrow{\qquad} O-C---O-C-C-(O-C-C)_{n+1}$$

Polymerization occurs by cleavage of the oxygen bond attached to the less-substituted carbon atom of the oxirane ring with inversion of the configuration of the secondary carbon atom and formation of a secondary alkoxide polymer molecule while the approaching monomer oxirane complexes with the now vacant aluminum atom site. Stereospecificity of the resulting polymer depends on whether the coordination of monomer is enantiomorphic selectic (97).

The polymerization of 2,3-epoxybutane with the same initiator as used with propylene oxide shows that the oxygen/substituted-carbon atom bond can be cleaved and, hence, a mechanism can be logically proposed to account for the head-to-head, tail-to-tail structures identified by Price and Vandenberg. With 2,3-epoxybutane, it was found that amorphous polymer could also be as pure disyndiotactic as the crystalline forms. Amorphous polymer could arise from short sequences of stereoregularity that were too short to form crystallizable segments. These could arise when coordination of monomer temporarily displaced alkoxide, interrupting chain growth, which, when resumed, could be selective for the antipode monomer.

Aluminum porphyrins (98) have been used to prepare polymers with very narrow molecular weight distribution and controlled number average molecular weight from propylene oxide without complicating side reactions (99). The aluminum porphyrins have a structure as described below, where X is Cl, OH, OR, or O_2CR and R is an alkyl group.

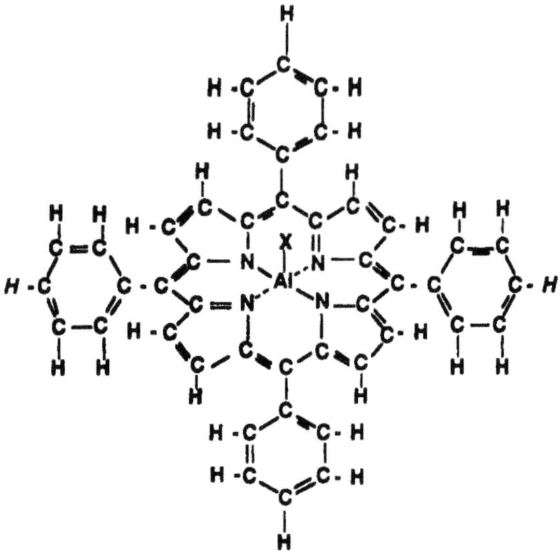

Aluminum porphyrin

Inoue and coworkers (100, 101) have examined these compounds for the polymerization of alkylene oxides, lactones, and lactides and termed them as initiators for immortal polymerization, in contrast to the living polymerization initiators (42).

When certain conditions regarding initiation are met (Section 4-III), living polymerizations yield narrow molecular weight distribution polymers, with the number of polymer molecules formed being essentially the same as the number of initiator molecules used. After exhaustion of monomer and polymerization ceases, more monomer or a different monomer can be added and polymerization again proceeds. Thus, the system is living. However, when a protic compound such as hydrogen chloride is added to the system, the propagating or living species reacts and terminates to yield a terminal hydroxyl group and an alkali metal chloride (102). As a result, the reaction is killed.

In the case of the immortal polymerization of alkylene oxides with aluminum porphyrins, the propagating species is also of a nucleophilic nature, and narrow molecular weight distribution polymers and copolymers are obtained even in the presence of protic compounds. Since the propagating or living species is not killed by protic compounds such as hydrogen chloride, alcohols, water, and carboxylic acids, it is said to be immortal. Another difference between immortal and living polymerizations is that with the former type, the number of polymer molecules is greater than the number of initiator molecules used.   NMR investigation (101) of the polymers indicated that the origin of the immortality lies in the unusual reactivity of the axial-group/ aluminum bond. For example, if the X group of the initiator is —OR and hydrochloric acid is added, there is a rapid, irreversible reaction that takes place at this site, with ROH forming and X becoming a chloride group. The new porphyrin remains capable of initiating polymerization.

Polymeric organometallic oxides have been used as initiators for the polymerization of ethylene oxide, propylene oxide, 1,2-butylene oxide, and epichlorohydrin (103). Poly($ArSbO_2$), where Ar is phenyl (Ph), p-methyl phenyl, and p-chlorophenyl, initiated polymerizations were 3 to 60 times faster than those obtained with compounds such as Ph_3SbO. NMR results indicated that the initiating species were $ArSbO_2$ or $Ar_2Sb_2O_4$ that were obtained through oxirane monomer solvation and subsequent fragmentation of the poly($ArSbO_2$).

A bimetal initiator system was explored by Kuntz (104) involving an aluminum complex similar to the Vandenberg Catalyst and diethylzinc. The bimetal complex could be envisaged as

$$
\begin{array}{c}
Et \diagdown \quad Et \\
Zn \\
\vdots \\
-O \diagdown \quad O \quad \diagup O- \\
Al \quad Al \\
-O \diagup \; R \; R \; \diagdown O-
\end{array}
$$

This complex, which was effective in initiating the homopolymerization of propylene oxide and epichlorohydrin and the copolymerization of these two monomers, was postulated to first form complex monomer with the zinc atom and then cleave by attack of the alkoxide linked to an aluminum atom to yield polymer.

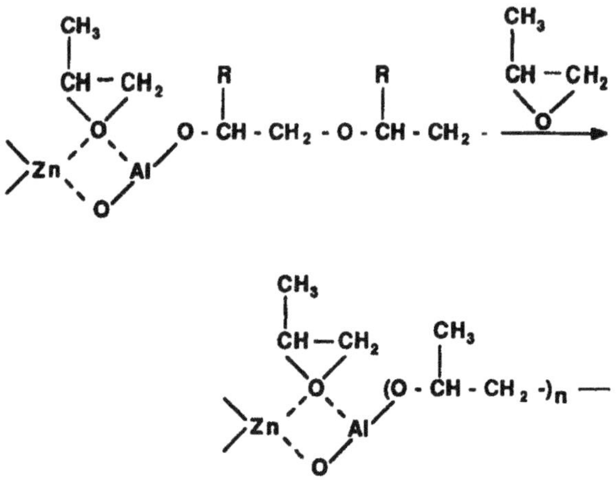

Inversion of configuration takes place with ring-opening carbon (103). Head-to-head polymerization can occur. Relative copolymerization rates depend on the relative coordinating ability of the comonomers, while copolymerization rate will depend on ring opening of the slower coordinating monomer (105).

Further insight into the polymerization of oxiranes can be found in the work of Tsuruta, using a well-defined organozinc complex (72, 96). The complex was prepared by reaction of 16 mmol of methanol with 14 mmol of diethylzinc in heptane. The complex

$$[Zn(OCH_3)_2 \cdot (C_2H_5ZnOCH_3)_6]$$

could be isolated as a crystal and its structure determined by X-ray diffraction. The complex is soluble in benzene without dissociation. In benzene, the complex is highly active toward initiation of stereospecific polymerization of propylene oxide at temperatures between 60°C and 80°C (Figure 3).

The complex structure of Tsuruta's organometallic complex, as determined by X-ray diffraction, consists of two enantiomorphic distorted cubes that share a common corner, zinc atom.

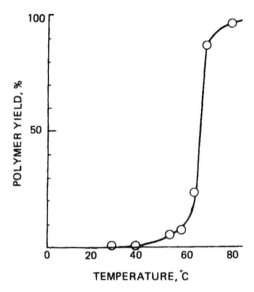

FIG. 3 Temperature dependence of the polymerization of propylene oxide with the Tsuruta catalyst, 4.8 mol/liter in benzene for 50 hours. (From Ref. 72.) Reprinted with permission of the copyright owner, Plenum Press, New York.

Monomer, propylene oxide, first coordinates with the central, shared zinc atom and an inner methoxide.

MeO —— ZnEt
EtZn —|—OMe   R
EtZn —|— OMe   O—CH
MeO —— Zn   CH₂-► OMe
     MeO ———— ZnEt
  MeO —|— ZnEt |
     EtZn ———— OMe

MeO —— ZnEt
EtZn—|—OMe   R   CH₃ OMe
EtZn —|— OMe   CH - CH₂
MeO —— Zn   O—CH
     CH₂◄-O
     MeO ——— ZnEt
  MeO —|— ZnEt |
     EtZn ———— OMe

The propylene oxide monomer unit that had formed a coordination bond with the central zinc atom undergoes ring opening by a nucleophilic attack of the neighboring methoxy group cleaving the ring at the oxygen-methylene bond. The polymer chain growth begins at this enantiomorphic slective site as another monomer unit reforms a complex with the central zinc atom. The mechanism is generally similar to that described with the Vandenberg Catalyst.

The complex is made up of two enantiomorphic, distorted cubes, one of which can be thought of as the "l" cube and the other as the "d" cube. The probability of either cube becoming the active site is equal, leading to an equal number of d-selective and l-selective sites generating stereoregular polymer. Carbon-13 NMR distinguishes between the methoxyls that are inner, near the central, shared zinc atom, and those that are outer, on the periphery of the complex. At temperatures near 60°C to 80°C, the spin-lattice constants of the inner methoxyls increase more rapidly than those of the outer methoxyls, suggesting that these inner methoxyls will become selectively more reactive. During polymerization, the increase in number of polymer terminal methoxy groups increases as the number of inner complex-methoxy groups decreases. These observations are in support of the proposed mechanism.

The polymerization kinetics were studied in detail. It appeared that the rate of reaction decreased as the concentration of complex increased. This anomalous behavior was explained in terms of the ability of all the zinc atoms in the complex to form coordination complexes with propylene oxide monomer, but only the inner methoxyls were effective in opening the oxirane ring of monomer coordinated with the central zinc atom. In support of this explanation, it is noted that neither diethylzinc alone nor ethylzinc methoxide initiate propylene oxide polymerization under the conditions described here; however, both will form complexes with propylene oxide, about two methyl oxiranes per zinc. Addition of either zinc alkyl to Tsuruta Catalyst/

propylene oxide markedly decreases the polymerization rate by reducing the effective monomer concentration in the system. Catalysts of this general nature have been used for initiation of the copolymerization of N-phenylmaleimide and propylene oxide (106).

Recently, a new crystalline organozinc enantiomorphic initiator for the stereoselective polymerization of propylene oxide was reported (107). The complex was obtained by reaction of diethylzinc and (dl)-1-methoxy-2-propanol, and it had the following composition:

$$\{[CH_3OCH_2CH(CH_3)O-Zn-OCH(CH_3)CH_2OCH_3)]_2 \cdot$$

$$[C_2H_5-Zn-OCH(CH_3)CH_2OCH_3]_2\}$$

The structure of this complex differed markedly from that of earlier reported organozinc complexes of this general type (96, 108). Zinc dialkoxide and ethyl zinc alkoxide were present in a 2:2 ratio rather than the 1:6 ratio of the other complexes, and they were in a centrally symmetric chair-form structure that consisted of bonded oxygen and zinc atoms. The chair structure was stabilized by coordination bonds with propoxy oxygen atoms. The complex is completely soluble and retains its chair structure in benzene solution as well as after initiation of polymerization. By considering that initiation took place through nucleophilic attack of a 1-methoxy-2-propoxy group at the methylene group of propylene oxide, and by measuring residual methoxypropoxy groups after polymerization, it was concluded that there was one active site on each molecule of complex. When optically active propylene oxide was used, it was possible to demonstrate that bond cleavage takes place in a highly regioselective manner, with over 90 percent of the cleavage occurring at the $O-CH_2$ linkage during propagation. Rate of polymerization was much greater with the chair-structure complex than with the distorted cube complexes, due to easier accessibility of monomer to the active sites, which presumably are located at the corners of the chair structure.

The most active polymerization initiators for the polymerization of substituted oxiranes, particularly propylene oxide, produce a significant amount of product that is stereospecific and crystallizable. Most uses for polymers of propylene oxide, however, are in elastomeric systems (see chapter 2, table 4). For elastomers, the amorphous, noncrystallizable polymers are pre-

ferred.   The principal commercial route to polymers has remained the stepwise anionic (base-catalyzed) processes.

Coordinate initiation routes to principally amorphous polymer
have been found that do offer certain advantages, although these
routes have been slow to be adopted, probably due to the apparently complicated structures of the initiators.   One of these
processes employs zinc with a complex anion promoted with a
low-molecular-weight polyether such as dioxane or diglyme (109).
The initiator system is based on zinc cobalticyanide (hexacyanocobaltate), $Zn_3[Co(CN)_6]_2$, or zinc ferricyanide (hexacyanofer-
rate), $Zn_3[Fe(CN)_6]_2$, (110–113).

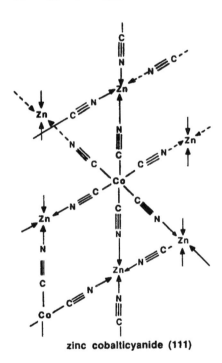

zinc cobalticyanide (111)

Zinc cobalticyanide can be obtained as an amorphous precipitate (114) from zinc chloride and potassium or calcium cobalticyanide in water.   The product has a formula

$$Zn_3[Co(CN)_6]_2 \cdot 12 \ H_2O$$

As prepared, the complex contains associated chloride ions,
which can be reduced with washing, and the associated water

can be reduced with drying from a formula value, as indicated, of about 12 to less than 1. Zinc cobalticyanide is a lattice that contains four-coordinate zinc ions and six-coordinate cobalt ions. As drawn above, the zinc cobalticyanide lattice would consist of sheets of cubes of cobalticyanide connected by zinc cations. This depicted structure would be too high in cobalt content in comparison to the actual initiator product. The initiator lattice has an occupancy of sites by zinc cations that is close to 100 percent; however, the cobalt occupancy is about 66 percent, which results in large lattice defects that can be occupied by water. In the depicted form, zinc cobalticyanide is not effective in initiating oxirane polymerization. Presumably, the zinc atoms required to coordinate oxirane are not available for interaction with monomer.

An initiator with improved activity is prepared by using dioxane or the dimethyl ether of ethylene glycol (glyme) or diethylene glycol (diglyme) as the reaction medium instead of water. The precipitated initiator recovered by filtration is reported to be largely amorphous and retains a significant proportion of ether in each case after washing and drying. It is generally observed that retained anions, which may be chloride or nitrate, depending on the zinc salt used, reduce initiator activity and may be found as end groups on some of the poly(alkylene oxide)s made. Zinc chloride, if present, is probably coordinated with water or alkyl ether in the defect regions. Water content of the initiator can be related to the molecular weight of the prepared poly(alkylene oxide), and it appears to act as a coinitiator/transfer agent. The initiator species can be described by the empirical formula:

$$Zn_a[Co(CN)_bX_c]_d(H_2O)_e(R)_f$$

where a, b, and d are numbers generally relating to the composition, which is basically zinc hexacyanocobaltate. The number c relates to anions, X, that have become occluded in the lattice during preparation and may be washed or leached out of the complex. The numbers e and f relate to the coordination number of zinc. Water remaining in the precipitated initiator after drying is denoted by e, which can be reduced to unity or less. The designator f relates to the amount of oxygenated activator present in the precipitate, usually an oxygen-containing, small organic molecule such as acetone, dioxane, or glyme.

The structure of the zinc hexacyanocobaltate complex in the vicinity of an active initiator site can then be pictured as

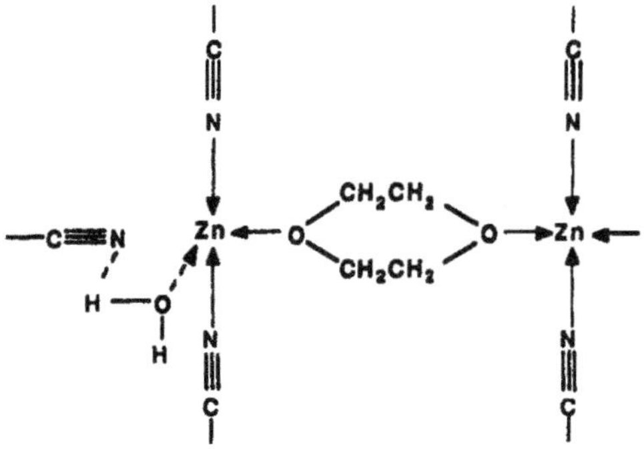

The activator polyether, in this case dioxane, opens-up the cy-
anide-zinc lattice in a manner that permits coordination with
oxirane, which displaces the activator polyether. This is followed by hydroxyl attack, in this case from water, and initiation of polymerization. Polymerization proceeds by rearward
attack of alkoxide on oxirane, so that the polyether produced
is terminated, in the case of propylene oxide, by secondary
hydroxyl groups.

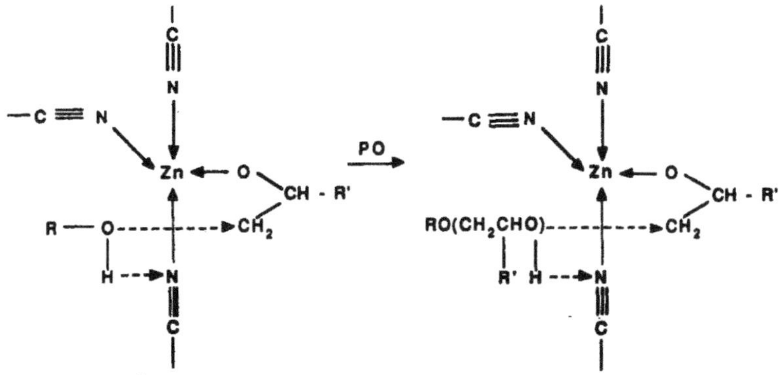

From this view, there is no clear way in which the initiator
site would be enantiomorphic selective, unless the activator polyether participated in site selectivity. The polymer formed with
propylene oxide and dioxane activator is believed to be largely
nonstereoregular.

Several features of these zinc-complex initiators make them
of commercial interest. The polyethers made from substituted
oxiranes are amorphous and, hence, suitable for elastomeric ap-

plications including polyurethane manufacture. There is little
tendency for chain termination by mechanisms described in Section II that lead to terminal unsaturation, an undesirable reduction in terminal hydroxyl functionality, which again makes the
products attractive for use in elastomers and polyurethanes.
Molecular weight control can be exercised by chain transfer reactions that may involve water, as in the formula suggested
above, or a hydroxyl or polyhydroxyl compound.

As a result of the chain transfer effects described, these
initiators can function as telomerization initiators (115). Telomerization is a form of polymerization in which one molecule reacts to both initiate and terminate, or limit, the degree of polymerization of monomer. Thus, a species, XY, would initiate
polymerization of a monomer, M, forming

$$X-M_n-$$

and then a part of a similar telogen, XY, would terminate the
polymer chain growth by forming

$$X-M_n-Y$$

In the case of these complex zinc hexacyanocobaltate initiators, the telogen could be a diol such as propylene glycol, a
triol such as glycerine, or a higher functionality polyol such as
pentaerythritol or sorbitol, and the product could then be a
polyol of the type used extensively in making polyurethanes in
elastomeric, foam, or other form.

In Table 4, following the organization procedure of Furukawa
and Saegusa, polymerization initiators for alkylene oxides were
divided into two groups: those that apparently were active
though metal-oxygen bonds and those based on alkaline earth
compounds. It is now possible to unify these two groups.
First, the alkaline earth carbonates and oxides (63, 64) are active only if a limited range of water concentration is associated
with the highly oxygenated surfaces of these heterogeneous initiators and if the surfaces are also free of occluded, strong
acid anions such as nitrate or sulfate. Initiator activity also
requires a conditioning, or induction, period in the presence of
oxirane (9).

Locally, near the initiator site, the arrangement of molecules
must fit the now familiar pattern of oxirane coordination followed
by rearward attack cleaving one oxygen-carbon bond. In the

case of the calcium carbonates, this can be represented by the
following structural equation.

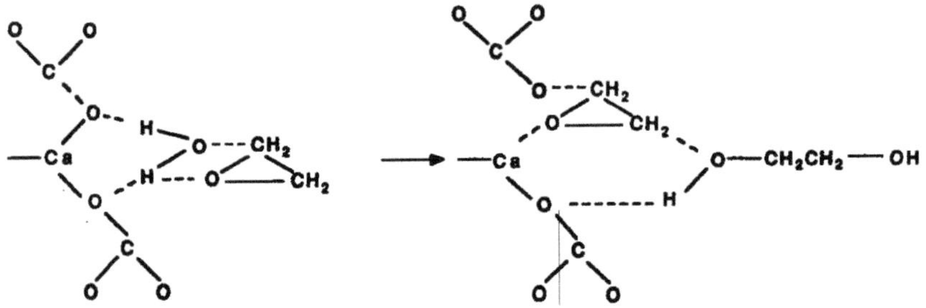

The amides are particularly effective when formed from
hexammoniate in the presence of oxide monomer (62–65). These
initiators, discovered by Bailey, Hill, and Fitzpatrick, can be
similarly depicted with the active initiator site,

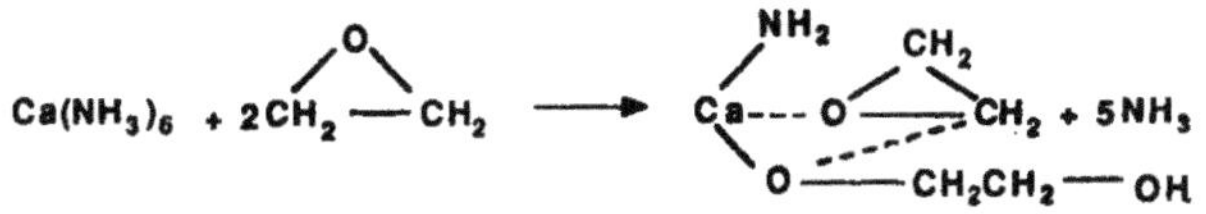

which proceeds to polymerize monomer, ethylene oxide, very
rapidly (116). These initiators, based principally on calcium
and magnesium, are much weaker than the aluminum or zinc
systems in forming oxirane complexes. As a result, they are
particularly effective in polymerizing ethylene oxide, which
forms complexes more readily in comparison with the substituted
oxiranes. With the substituted oxiranes, propylene oxide in
particular, the calcium initiators of Bailey et al. form weaker
complexes and initiate polymerization at a slower rate. The polymers formed are primarily the amorphous polyethers.

Lamot (117) prepared novel, non-nitrogen-containing compounds from calcium and diphenylethylene, $(Ar)_2Ca$, and then
complexed the compounds with various amounts of epichlorohydrin, ECl, with tetrahydrofuran, or with a 1:0.7 mole ratio of
epichlorohydrin to tetrahydrofuran. After preliminary testing
of the complexes as initiators for polymerization of epichlorohy-

drin, an $(Ar)_2Ca \cdot O \cdot 7ECl$ complex was chosen as the initiator for a detailed study of polymerization under a variety of conditions.

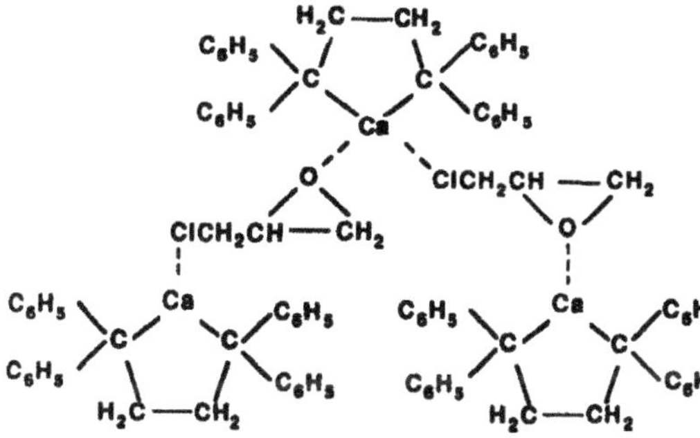

$(Ar)_2Ca \cdot O \cdot 7ECl$ Complex

With this complex, partially crystalline, soluble polyepichlorohydrin with viscosity-average molecular weights greater than 2×10^5 and yields greater than 70 percent were obtained.

A highly branched or star-shaped polyoxyethylene has been prepared by using divinylbenzene to obtain a multicarbanionic core to which ethylene oxide was added (118). The early stages of polymerization took place in a heterogeneous environment, but this did not interfere with the main propagation step, and a star with many arms was obtained. The swelling behavior of divinylbenzene-styrene-ethylene oxide graft copolymers has been investigated by means of equilibrium swelling and uniaxial compression measurements in the swollen state (119). Even though there is incompatibility between the polystyrene network and the polyoxyethylene, the grafts had only a small effect on the network swelling behavior. The compression modulus was essentially independent of the grafted polymer chains.

Potassium dissolved in tetrahydrofuran containing 18-crown-6 ether was used as an initiator for the polymerization of ethylene oxide (120). The number of active centers in the polymerization was shown to be equal to the initial 18-crown-6 concentration.

V.   COPOLYMERIZATION OF ALKYLENE OXIDES

Copolymerization of alkylene oxides is complicated because the
results depend on the initiator system; the solvent, if present;
and the terminal or end group (and sometimes the penultimate
group as well). The simplifying factor is that the copolymerization kinetics can be characterized by a single relative reactivity rate ratio (121) rather than the two relative reactivity
ratios required for free-radical, vinyl copolymerization (122).
There are very few reported instances of free-radical copolymerization of oxirane (123).

In cationic initiation, true copolymerization is difficult due
to the wide range of reactivity of oxiranes with acid or cationic
attack (Table 1). Generally, the copolymer formed is lower in
molecular weight, with a significant amount of cyclic product
formed, than is the case in homopolymerization. Except at very
low temperatures, around −78°C, linear copolymers tend to terminate by proton transfer before significant molecular weights
are achieved when cationic initiation is used.

In coordinate-initiated polymerization, certain initiators are
able to function by either a coordinate or cationic mechanism
(88, 124). In attempts to use these initiators in copolymerization with monomers of widely differing reactivity toward cationic
ring opening, the polymer obtained is predominantly one monomer, although block polymers may be formed.

Anionic, stepwise polymerization is used to prepare copolymers of alkylene oxides. The relative copolymerization rate is
essentially 1. Random or block polymers can be formed, depending on the monomer fed to the reaction.

The mechanism of polymerization is the same as that discussed in Section III. The initiating anion, RO^-, is usually a
complex with an alkanol that can be a monol, a diol such as
ethylene or propylene glycol, a triol such as glycerine, or a
higher functionality polyol such as pentaerythritol or a sugar
such as sorbitol or sucrose. The relative copolymerization rate
of alkylene oxides is close to unity; however, the rate of reaction depends on the specific monomer combination. The rate-
determining step is the ring-opening reaction of the slower
monomer:

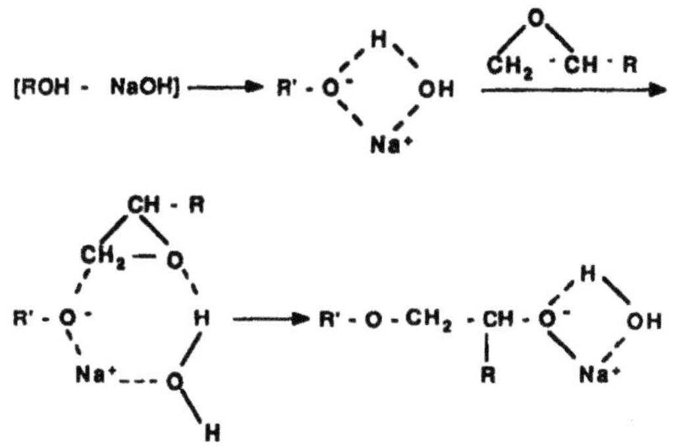

where R can be hydrogen, methyl, ethyl, or higher alkyl (ethylene oxide, propylene oxide, 1,2-butylene oxide, etc.) and R'OH is the starting alkanol such as allyl alcohol, n-butanol, glycol, glycerine, etc. In principle, but within the reactivity limits indicated in Table 1, higher substituted oxiranes can also copolymerize.

The most common initiators used are sodium or potassium hydroxide together with the starter alcohol. There is some preference for potassium hydroxide because, if relatively high molecular weights are desired and propylene oxide is one of the monomers in the reaction, there is less tendency to terminate a reaction chain by hydride transfer giving terminal unsaturation with potassium than with sodium. Indeed, it is reported that cesium gives a still lower level of termination and terminal unsaturation (Section III).

As examples, polymers of ethylene oxide can be formed from an ethylene glycol starter

$$n \; CH_2\!-\!CH_2 + HO\!-\!CH_2\!-\!CH_2\!-\!OH \; -\!KOH\!\rightarrow \; HO\!-\!(CH_2\!-\!CH_2\!-\!O)_{n+1}H$$

    molecular   molecular weight

    weight = 44   ~200—20,000

or of propylene oxide from a propylene glycol starter

$$n \ CH_2-CH \overset{CH_3}{\underset{O}{|}} + HO-CH_2CH-OH \overset{CH_3}{\underset{}{|}} \ -KOH\rightarrow \ HO-(CH_2CH-O)_{n+1}H \overset{CH_3}{\underset{}{|}}$$

molecular
weight = 58

molecular weight
~400—10,000

and copolymers can be formed from either starter if a mixed
feed of ethylene oxide and propylene oxide is used.

$$-CH_2CH_2OCH_2CH_2OCH_2CHOCH_2CH_2OCH_2CHOCH_2CHOCH_2CHOCH_2CHO-$$
$$\underset{CH_3}{|} \qquad \underset{CH_3}{|} \quad \underset{CH_3}{|} \qquad \underset{CH_3}{|}$$

A random copolymer of ethylene oxide and propylene oxide,
the composition of which depends on the relative concentrations of monomer present

Block copolymer can be prepared from either starter by alternately feeding first one and then the other monomer.

$$-CH_2CH_2OCH_2CH_2OCH_2CH_2OCH_2CHOCH_2CHOCH_2CHOCH_2CHO-$$
$$\underset{CH_3}{|} \quad \underset{CH_3}{|} \quad \underset{CH_3}{|} \quad \underset{CH_3}{|}$$

A block copolymer of ethylene oxide and propylene oxide,
the composition of which depends on the time each monomer
is present during the reaction

Since these anionic stepwise reactions are essentially living polymerizations (42, 43), complete reaction of one monomer can be
followed, sequentially, by reaction of a second, giving rise to
block polymers with AB or ABA structures. It should also be
apparent that block copolymers with random block compositions
of different monomer ratios can be prepared.

The starting alcohol can be a higher alkanol, such as do-
decanol or nonylphenol, or a phenoxide, which can also initiate
polymerization. In addition, the starting group can be other
than a hydroxyl group such as an amine, a diamine, or an
amide. Although these different polymerization starters may
seem to be insignificant relative to the alkylene oxide portion of
the molecule, they become particularly important in the design
of surface-active materials and compounds used in special preparations.

In the preparation of polyols for urethane polymer intermediates, higher functionality polyols are usually required. The most common starter is glycerine to produce triols.

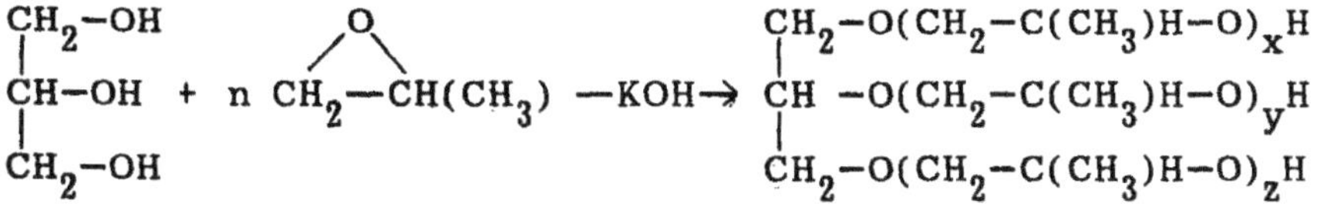

$$\begin{array}{l} CH_2-OH \\ | \\ CH-OH \\ | \\ CH_2-OH \end{array} + \; n \; CH_2\overset{O}{\diagdown}CH(CH_3) \; -KOH \rightarrow \begin{array}{l} CH_2-O(CH_2-C(CH_3)H-O)_xH \\ | \\ CH \; -O(CH_2-C(CH_3)H-O)_yH \\ | \\ CH_2-O(CH_2-C(CH_3)H-O)_zH \end{array}$$

where n = x + y + z

The triol produced may also be a random copolymer, a block copolymer, or a block copolymer in which only a small percentage of the second monomer is used at the end of the reaction to modify the terminal groups. Commonly, a small amount of ethylene oxide is used to convert the terminal hydroxyl groups from secondary hydroxyl groups to primary hydroxyl groups to take advantage of the higher reactivity of primary hydroxyls in subsequent reactions, such as the formation of urethane in the preparation of polyurethanes.

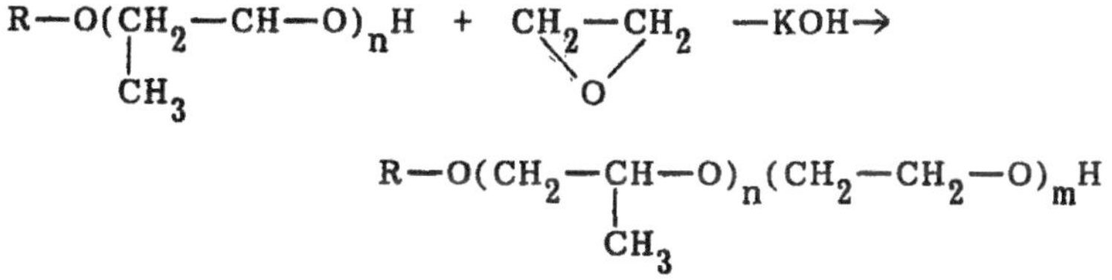

$$R-O(\underset{\underset{CH_3}{|}}{CH_2}-CH-O)_nH \; + \; CH_2\overset{\diagup CH_2}{\underset{O}{\diagdown\diagup}} \; -KOH \rightarrow$$

$$R-O(CH_2-\underset{\underset{CH_3}{|}}{CH}-O)_n(CH_2-CH_2-O)_mH$$

$$n \gg m$$

A cyclic anhydride, such as phthalic anhydride or maleic anhydride, will form a nearly 1:1 alternating copolymer with alkylene oxides. This reaction is the same one used in the curing (hardening) of epoxy resins. A large number of initiators can be used for this reaction, including tertiary-amines, various Lewis acids, and bases including alkoxides. The uncatalyzed reaction invokes initiation with a hydroxyl group (125, 126). This reaction can be used to introduce a polyester unit into a polyether polymer chain. With phthalic anhydride, the reaction proceeds as follows:

$$RO^- \cdot Na^+ + O = C \underset{\substack{C-C \\ C \\ C = C}}{\overset{O}{\underset{}{\bigcirc}}} C = O + CH_2 - CH \cdot R \longrightarrow RO \cdot C \underset{\substack{C-C \\ C \\ C = C}}{\overset{O \quad O}{\underset{}{\bigcirc}}} C \cdot O \cdot CH_2 CH \cdot O^- \cdot Na^+$$

The rate of this reaction is proportional to the concentration of alkoxide, epoxide, and anhydride.

Alkylene oxide-derived glycols are used to prepare unsaturated polyester intermediates from unsaturated and saturated anhydrides for subsequent reaction with styrene or other suitable ethylenically unsaturated monomers. Such products are used in the preparation of molded parts such as automobile hoods, business machine housings, and so on. Ethylene and propylene glycols are used to decrease costs and to improve water resistance, strength, and heat resistance. Diethylene and dipropylene glycols are used for similar reasons and also for improved flexibility and toughness in the molded or cast part. Although these are low-cost intermediates, it is more cost effective on a mole basis and a production-time basis to use the corresponding alkylene oxides. The alternating polymerization mechanism somewhat limits the degree of flexibility and toughness that can be built into the polyester and the final manufactured part. However, it has been demonstrated that zinc chloride can promote both the esterification and etherification reactions (127). For maleic anhydride/propylene oxide copolymers, it has been found that organozinc compounds can be used to alter the amount of propylene oxide incorporated into the molecules (128, 129). The greater the acidity of the zinc, the greater the amount of propylene oxide incorporated and the less the degree of alternation. In at least once instance, propylene oxide has been used in combination with maleic and phthalic anhydrides and styrene in a one-step process to produce cast parts (130). The parts produced by this process were inferior to those produced in a conventional manner.

A cyclic anhydride, such as succinic, phthalic, or maleic anhydride, can also be used to "tip" or convert the terminal hydroxyl groups to carboxylates:

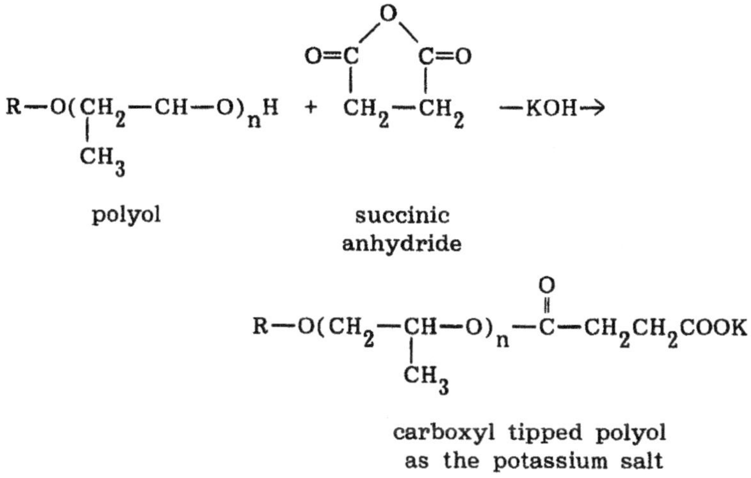

$$O=C \overset{O}{\diagdown} C=O$$

R—O(CH$_2$—CH—O)$_n$H + CH$_2$—CH$_2$ —KOH→
　　　　|
　　　CH$_3$

polyol   succinic
                                    anhydride

$$R—O(CH_2—CH—O)_n—\overset{\overset{O}{\|}}{C}—CH_2CH_2COOK$$
　　　　　　　|
　　　　　　CH$_3$

carboxyl tipped polyol
as the potassium salt

Such carboxylates have been used to prepare segmented block copolymers from ethylene oxide and beta-pivalolactone (131—133). Ethylene oxide is first polymerized with an alkoxide initiator, and the resulting polyoxyethylene alkoxide is modified with succinic anhydride to yield a carboxylate anion. The anion is then used to polymerize beta-pivalolactone as a polymeric block. After treatment with hydrogen chloride, a succinate-linked AB block copolymer with a terminal hydroxyl group on the polyoxyethylene block and a terminal carboxyl group on the polylactone block results. The AB block copolymer is then coupled via a condensation reaction to prepare the (AB)$_n$ segmented block copolymer.

Rhinocladiella atrovirens KY 801 will readily cause microbial transformation of the terminal hydroxyl end groups of polyethylene glycol mono- and dicarboxylic compounds (134). The process does not have the drawbacks of catalytic oxidation—which, in addition to carboxylating the molecules, causes degraditive oxidation of the main backbone and aldehyde formation. Polyethylene glycols with molecular weights of up to 2000 were converted to the corresponding carboxylic acid-terminated molecules.

Cyclic carbonates of glycols—e.g., ethylene carbonate or propylene carbonate—are readily prepared from carbon dioxide and the alkylene oxide (135).

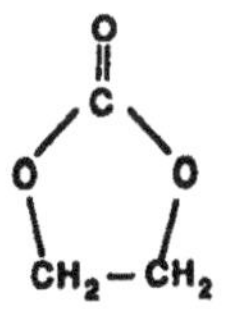

Ethylene Carbonate 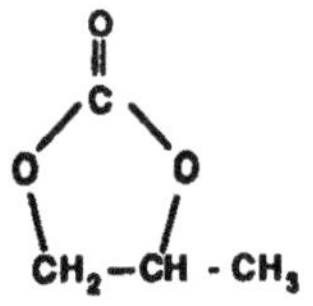   Propylene Carbonate

The cyclic carbonate can then be copolymerized with cyclic anhydride to yield, with loss of carbon dioxide, a polyester of
relatively high molecular weight. This copolymerization is initiated by a number of inorganic salts including sodium cyanide
and potassium chloride, bases such as potassium hydroxide or
sodium amide, and Lewis acids at 200°C. Presumably, the initiation begins with nucleophilic attack of an anion on the cyclic
carbonate

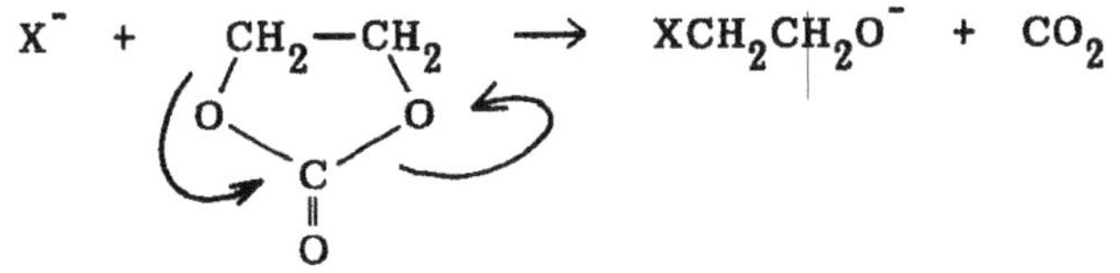

followed by opening of the cyclic anhydride ring, leading to an
alternating copolymer, a polyester of the alkylene oxide, and
the acid anhydride.

Recently, new high-molecular-weight poly(alkylene carbonates) formed by reaction of either ethylene or propylene oxide
with carbon dioxide were introduced to the marketplace (136).

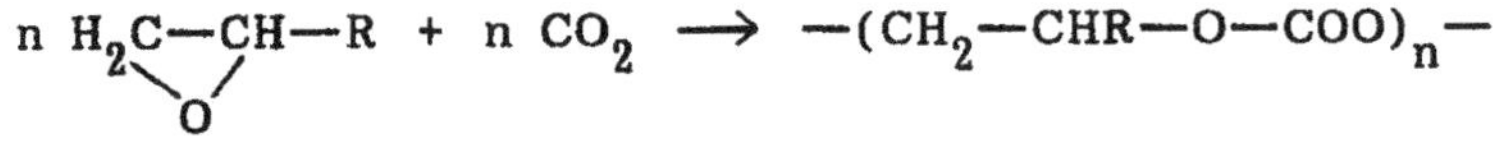

$$n\ H_2C-CH-R\ +\ n\ CO_2 \longrightarrow -(CH_2-CHR-O-COO)_n-$$

These polymers, with number average molecular weights of about
50,000, have decomposition temperatures of 220–250°C and good
to excellent tensile properties. The poly(alkylene carbonates)
decompose and burn cleanly to yield mainly the appropriate cyclic carbonate with minimal ash residue and only trace levels of
contaminants. Because of these clean burning characteristics,
suggested applications include use as a ceramic binder metal-
casting material.

The discovery of coordinate polymerization of oxirane led to an enormous expansion in the range of polyether structures that could be synthesized. Some of the remarkable synthesis chemistry is outlined in Section III. Coordinate copolymerization further increased the range of polymer structures that could be synthesized in highly controlled ways. Two examples have already been described: copolymers of epichlorohydrin and ethylene oxide (71) and of propylene oxide and allyl glycidyl ether (94, 137). The activation energy for copolymerization of epichlorohydrin and maleic anhydride was found to be 14.5 kcal/mol, and the polymerization rate is dependent on temperature and proportional to catalyst concentration (138–140).

Some of the complexities of coordination copolymerization of alkylene oxides are highlighted by the work of Vandenberg with 2,3-epoxybutanes (88). Using the Vandenberg Catalyst, it was shown that one could selectively polymerize cis-2,3-epoxybutane from an equal mixture of cis- and trans-isomers and obtain fairly pure cis-polymer. The cis-oxide enters the crystalline polymer fraction about 20 times faster than does the trans-oxide. In the amorphous fraction of the polymer produced, the enrichment of cis-oxide was about tenfold over that of the trans-isomer.

At low temperatures, about −78°C, in triisobutylaluminum/water-initiated polymerization, which is presumed to be a cationic initiation, an amorphous (elastomeric) polymer is obtained from cis-2,3-epoxybutane, and a crystalline polymer, melting point 100°C, is obtained from the trans-isomer. In a copolymerization of the two isomers, the cis-oxide enters the copolymer at about twice the rate of the trans-isomer. Further, the low-temperature, cationic poly(trans-2,3-epoxybutane) with a crystalline melting point of 100°C was found to consist of diad units with a mesodiisotactic structure, while the crystalline polymer formed by coordinate polymerization of the cis-monomer, melting point 162°C, had diad units that were racemic diisotactic. These results make apparent the importance of the monomer coordination step in polymer chain growth in coordinate polymerizations.

Bailey and France (104) proposed that with a range of coordination initiators, copolymerization propagation consisted of two steps: coordination of monomer to an active site, metal or hydrogen, followed by rearrangement during addition with alkoxide and regeneration of an alkoxide ion.

$$RO^- \quad X^+ \;+\; \underset{R_2}{\overset{R_1}{C}}\!\!-\!\!O\!\!-\!\!\underset{R_3}{\overset{R_4}{C}} \;\rightleftharpoons\; RO^- \quad \underset{R_2}{\overset{R_1}{C}}\!\!\overset{X^+}{-O-}\!\!\underset{R_3}{\overset{R_4}{C}}$$

$$RO^- \quad \underset{R_2}{\overset{R_1}{C}}\!\!\overset{X^+}{-O-}\!\!\underset{R_3}{\overset{R_4}{C}} \;\longrightarrow\; RO-\underset{R_2}{\overset{R_1}{C}}-\underset{R_3}{\overset{R_4}{C}}-O^- \quad X^+$$

In the first step, the equilibrium of coordination should shift toward coordination with increasing degree of substitution of monomer. However, steric effects resulting from increased substitution should reduce the rate constant for rearrangement leading to chain growth. This explanation was based on the copolymerization behavior of a range of alkylene oxides with dibutylzinc, calcium glycoxide, calcium amide, and zinc carbonate initiators (Tables 10 and 11).

Vogl and coworkers (4) copolymerized omega-epoxyalkanoates with a range of alkylene oxides using the Vandenberg Catalyst and obtained comparable results. With a range of alkylene oxides and one coordination initiator, relative copolymerization rates indicated that random copolymers were formed (Table 12). The copolymerization of ethylene oxide and cyclohexene oxide with the Vandenberg Catalyst also proceeds with a relative copolymerization rate of about 1.0 (88).

Initiators that have a particularly high affinity for coordinating specific alkylene oxides shift the relative copolymerization rate. These shifts depend on the steric volume in proximity to the coordination site, especially with bulky monomers such as isobutylene oxide. If special care is taken to prevent cationic polymerization, which will produce exclusively lower molecular weight polymers of isobutylene oxide, high polymer via a coordinate mechanism can be obtained with magnesium initiators such as $R_2Mg \cdot NH_3$ and amine-modified R_2Zn/H_2O initiators (141, 142). Crystalline poly(isobutylene oxide) is an entirely head-to-tail polymer structure with a crystalline melting point of about 155°C, entirely different in structure from the amorphous polymer obtained through cationic initiation.

Speranza and coworkers (143) modified cationically initiated ethylene oxide/propylene oxide copolymers by coreacting them, during the alkylene oxide copolymerization, with the diglycidyl

TABLE 10 Rate of Coordinate Copolymerization of Alkylene Oxides (From Ref. 105)

Alkylene oxide	Reactivity
R_1, R_2, R_3, R_4 = H (ethylene oxide)	Highest
R_1, R_2, R_3 = H (propylene oxide, styrene oxide)	
R_2, R_3 = ring system and R_1, R_4 = H (vinylcyclohexene monoxide)	
R_1, R_4 = alkyl and R_2, R_3 = H (2,3-epoxybutane)	
R_1, R_2 = alkyl and R_3, R_4 = H (isobutylene oxide)	Least

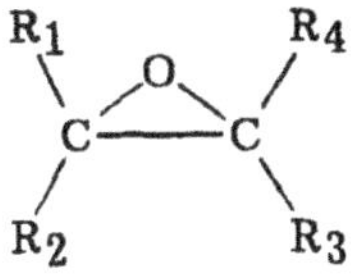

The structure shown at the top of the table:

$$R_1, R_2 \overset{O}{\underset{}{C-C}} R_4, R_3$$

ether of bisphenol A in such a manner that the diepoxide was introduced at selected points along the length of the polyol molecular chain. This modification tended to increase the overall average hydroxyl functionality of the polyols; when they were used to make polyurethane foams, the resulting products had higher load-bearing properties than did foams made with unmodified polyols.

High-molecular-weight copolymers of ethylene oxide and large alkyl chain 1,2-epoxides were prepared by Landoll (144) using a modified trialkylaluminum initiator system. Incorporation of a few percent or less of 14 to 24 carbon atom alkyl-1,2-epoxides into the polyoxyethylene chain resulted in copolymers with molecular weights of 300,000 to 700,000 that had improved

TABLE 11   Relative Copolymerization Rate of Alkylene Oxides

Monomer 1	Monomer 2	Initiator*	Relative co-polymerization rate[†]
Ethylene oxide	Propylene oxide	1,2,3,4	1.3
Ethylene oxide	Allyl glycidyl ether	2,4	1.5
Ethylene oxide	Styrene oxide	1,2,4	1.0
Ethylene oxide	4-vinylcyclo-hexene dioxide	2	1.0
Ethylene oxide	Butadiene monoxide	2,4	3.0
Ethylene oxide	Dicyclopen-tene dioxide	2	1—2
Ethylene oxide	Vinylcyclo-hexene monoxide	4	1.5—2.0
Propylene oxide	Styrene oxide	1	1.0
Propylene oxide	Allyl glycidyl ether	1,2	1.0

*Initiator: (1) dibutylzinc, (2) calcium glycoxide, (3) calcium amide, (4) zinc carbonate
[†]Relative copolymerization rate using the single ratio (122); i.e., the product of the reactivity ratios, $r_1 \cdot r_2$, is about 1 and the copolymer appears to be a random copolymer.

viscosity stability over that of the corresponding homopolymers. In addition, the copolymers interact with ethoxylated nonionic surfactants to form very high-viscosity solutions and even elastic, self-healing gels in certain instances.   For example, a 1 percent aqueous solution of one of the copolymers had a viscosity of 91 cP.   When a 10-mole ethylene oxide adduct of octyl phenol was added, the viscosity increased to 148,000 cP.

    A number of investigators have studied the copolymerization of ethylene oxide and propylene oxide with epsilon-caprolactone using a variety of active hydrogen starters such as water, 1,4-

TABLE 12 Copolymerization of Alkylene Oxides with Omega-Epoxyalkanoates with AlEt$_3$/H$_2$O/AcAc Initiator (Vandenberg Catalyst) (From Ref. 4.)

Comonomer	Epoxyalkanoate		
	Percent of Epoxyalkanoate Found in the Polymer Formed from a Monomer Mix Containing 30% Epoxyalkanoate		
	Methyl 4,5-Epoxypent-anaote	Methyl 7,8-Epoxyoct-anoate	Methyl 10,11-Epoxyundec-anoate
Ethylene oxide	—	30	15
Propylene oxide	27	30	20
1-butene oxide	25	30	25
1-hexene oxide	25	30	25
Phenyl glycidyl ether	25	30	25

butanediol, and neopentyl glycol and anionic, cationic, and co-ordination initiators (145–152). The compounds were prepared for use as poly(vinyl chloride) plasticizers and soft segments for synthesis of polyurethane foams (153, 154), elastomers (155), and press-molded articles (156). Brochet and coworkers (157) used DTA (differential thermal analysis) and NMR to investigate the structure of ethylene oxide/epsilon-caprolactone copolymers prepared with anionic and cationic initiators. Their results indicated that the copolymers prepared by anionic initiation were terminated with ethylene oxide segments. In the case of cationic initiation, the polymer molecules were made up of short chain segments of each monomer in a predominantly alternating structure—i.e., an (AB)$_n$-type block copolymer.

Propylene oxide has also been copolymerized with beta-propiolactone, epichlorohydrin, and methyl methacrylate using the alkaline earth/aluminum compound-type initiators (151). Terpolymers from ethylene oxide or propylene oxide, epichlorohydrin, and epsilon-caprolactone, as well as from epichlorhydrin, epsilon-caprolactone, and 2,2-bis(chloromethyl)oxetane,

have been made (158). Conversions varied from 80 to 100 percent when triisobutylaluminum/zinc acetylacetonate/water coordination initiators were used. Incremental addition of the epsilon-
caprolactone increased conversion from 73 percent to 94 percent.
Terpolymers for use in radiation-curable coating systems were
prepared from propylene oxide, epsilon-caprolactone, and 2-
hydroxyethyl acrylate (159).

VI. BLOCK COPOLYMERS OF ALKYLENE OXIDES

Block copolymers of alkylene oxides were introduced as
PLURONIC^R polyols by Wyandotte, now BASF Wyandotte Corporation, in 1951. The first products were made by alkoxylation of propylene glycol with propylene oxide to a molecular
weight of about 900, followed by further reaction with ethylene
oxide. These products can be generally characterized by the
formula

$$H(OCH_2CH_2)_a(OCHCH_2)_b(OCH_2CH_2)_cOH$$
$$| $$
$$CH_3$$

where b > 15 and the quantity b/(a + c) is in the range of
0.08 to 3.0. The products may be liquids, pastes, or solids
depending on composition and molecular weight. They can be
diols, as indicated above, starting the propylene oxide polyol
with propylene glycol, but they can also be monols or polyols
with a hydroxyl functionality of 3 or more.

Hydrophobic blocks are more often poly(propylene oxide)
or poly(1,2-butylene oxide); however, epichlorohydrin, tetra-
hydrofurans, styrene oxide, cyclohexene oxide, and allyl
glycidyl ether can be used. The hydrophilic blocks are generally poly(ethylene oxide), although glycidol and butadiene mon-
oxide can be used.

A further variation in structure can be achieved using diamines such as ethylenediamine or 2,4-toluenediamine. Actually,
any starter molecule can be used as a basis for block copolymer, provided the starter molecule contains a labile hydrogen
that will react to open a 1,2-epoxide. Usually, such labile hy-
drogens are defined as being reactive with methylmagnesium
chloride to liberate hydrogen in the Zerewitinoff Reaction (160).

With ethylenediamine, a surface-active product can be obtained
that is tetrafunctional (a tetrol) by first reacting the amine
with propylene oxide and then capping with ethylene oxide,

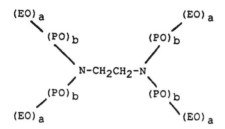

where a and b are average degrees of alkoxylation and usually
none of the a's or b's have the same value.

As will be discussed in Chapter 6, which deals with the
physical properties of the poly(alkylene oxide)s, only poly(ethylene oxide) is generally considered to be water-soluble; however, ethylene oxide polymers are water-soluble only within a
specific temperature range. There are, in theory, upper and
lower consolute temperatures, and for poly(ethylene oxide) in
water, the "Flory" or "theta" temperature is 108°C. There are,
also theoretically, consolute temperatures for other lower poly-
(alkylene oxide)s in water, though these temperatures may be
outside the range of normally liquid water. As a result, a hydrophobe can be achieved by forming a random copolymer of
ethylene oxide with another epoxide; within a temperature and
composition range, this copolymer block can serve as the hydrophobic block. Such a unit is sometimes termed a heteric
block, and the block copolymer may consist of blocks of the
heteric block and of poly(ethylene oxide).

As indicated above, an extraordinary range of compositions
can be synthesized in a relatively simple way that will encom-
pass surfactant properties that depend on composition and use
temperature. Virtually no other polymer system can span this
range of interfacial or surfactant properties. The polymerization mechanisms and procedures used are described earlier in
this chapter; and though most often anionic-step polymerization
is used, it is possible to prepare the block copolymers by cationic polymerization.

Generally, degree of polymerization is measured either by
product viscosity, usually either a Brookfield viscosity at 25°C
or a cup viscosity at 100°F, or by product hydroxyl number.
Hydroxyl number is defined as

$$\text{OH No.} = \frac{56100 \times f}{M_n} = \text{mg KOH/g of sample,}$$

where f is the hydroxyl functionality and M_n is the number average molecular weight. When the functionality is unknown, the highly useful parameter of equivalent weight, number average molecular weight divided by functionality, is determined. (Note: The constant 56,100 is a summary of the many parameters including the molecular weight of potassium hydroxide, 56.1, that yield the final units.) The hydroxyl number is determined by first forming the monophthalate ester through reaction of the polyol with phthalic anhydride and then back-titrating with potassium hydroxide to determine the amount of monophthalate formed. The method assumes that the acid number of the sample is small and negligible. However, if this number is significant, it must be taken into account, since it will affect the final result and decrease the measured hydroxyl number, which will, in turn, increase the apparent calculated equivalent weight or molecular weight (see Chapter 5).

The products produced are devolatilized by vacuum stripping at temperatures around 160°C or by stripping while sparging with an inert gas, such as dry nitrogen, to remove unreacted epoxide and low-molecular-weight products, particularly cyclic dimers and trimers. Catalyst (initiator) residues present are neutralized with acid (in base-initiated polymerizations), often phosphoric acid, or by ion exchange (161). The product can also be decolorized by passing it through beds of charcoal or clays such as magnasol (162).

Polyethers, poly(alkylene oxides), generally must be stabilized toward oxidation. Antioxidants such as catechol, t-butyl hydroxytoluene, or monoethers of hydroquinone are added (163). Phenothiazine is often used to stabilize against ultraviolet degradation, and various alkanolamines are used as thermal stabilizers (164, 165).

In Section III, polymerization of 1,2-epoxides by alkoxylation (stepwise anionic polymerization) was described using as starters a variety of lower molecular weight alcohols that may be monols such as methanol or lauryl alcohol, diols such as propylene glycol or tripropylene glycol, triols such as glycerol or trimethylol propane, or higher functionality polyols such as pentaerythritol or sorbitol. The processes used are essentially the same. The products differ in molecular weight, composition, and hydroxyl functionality. Depending on composition, hydrophobic-hydrophilic balance, and molecular weight, the products find uses in applications ranging from nonionic surfactants (166) to intermediates for urethane elastomers, coatings, molded plastics, and foams (167).

It should be apparent that block copolymers can be prepared from hydroxyl-terminated polyols of various types by reaction with diisocyanates or other difunctional compounds that react with hydroxyl groups such as epoxides, carboxylic acids, anhydrides, and so on. For example, if a difunctional, hydroxyl-terminated propylene oxide polyol molecule is end-capped with two molecules of a diisocyanate, leaving a free isocyanate group on each end of the molecule, it could be further reacted with two molecules of a monofunctional ethylene oxide polymer or oligomer. Such a reaction sequence would form an ABA, polyoxyethylene/polyoxypropylene/polyoxyethylene, block copolymer. A dihydroxyl functional polyoxyethylene could have been used in the above example, but product reproducibility would be more difficult, and the block distribution is far more complex than in the idealized example. However, this concept can be built on to make a variety of products from the mentioned polyols or other hydroxyl-containing oligomers or polymers. Functionalities greater than 2 can be used, but the results are far more complex, with gellation and/or insolubility readily occurring.

VII. BLOCK COPOLYMERS OF ALKYLENE OXIDES WITH STYRENE AND LACTONES

Advantage can be taken of the living polymer principle to prepare a wide range of highly controlled-structure polymers that are amphiphilic. The copolymers of this type that have been studied most are block and graft copolymers of styrene and alkylene oxides. These polymers can form amphiphilic micellar aggregates (168) and polymers with unique phase-transfer activity (169).

The simplest structures, which can be either block or triblock copolymers, can be obtained by sequential polymerization of styrene and ethylene oxide. If styrene is first polymerized with an alkali metal alkyl initiator, butyllithium or amylsodium ((42, 170, 171),

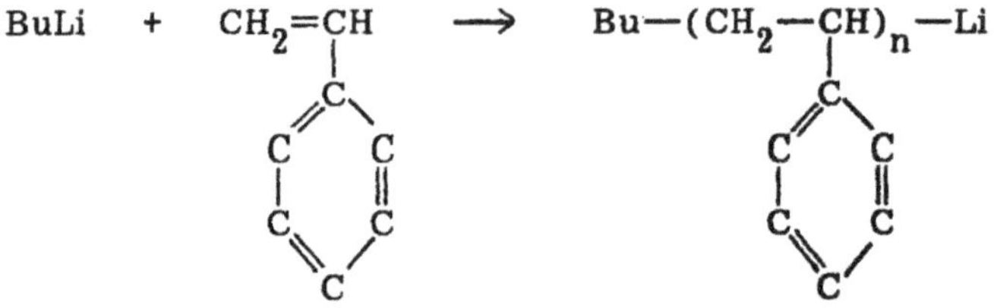

$$BuLi \ + \ CH_2{=}CH(C_6H_5) \ \longrightarrow \ Bu{-}(CH_2{-}CH)_n{-}Li$$

When the desired chain length of polystyrylithium is achieved, addition of ethylene oxide will terminate the polystyryl chains with ethoxide, which in turn can initiate polymerization of ethylene oxide to produce an AB block copolymer.

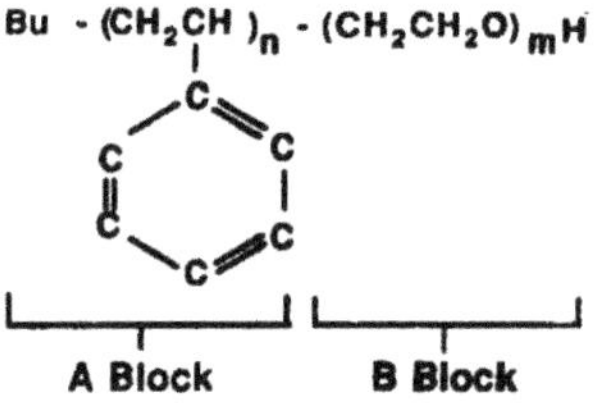

$$Bu\text{-}(CH_2CH)_n\text{-}(CH_2CH_2O)_mH$$

A Block   B Block

A triblock, ABA, copolymer can be obtained with a sodium naphthalene initiator (172).

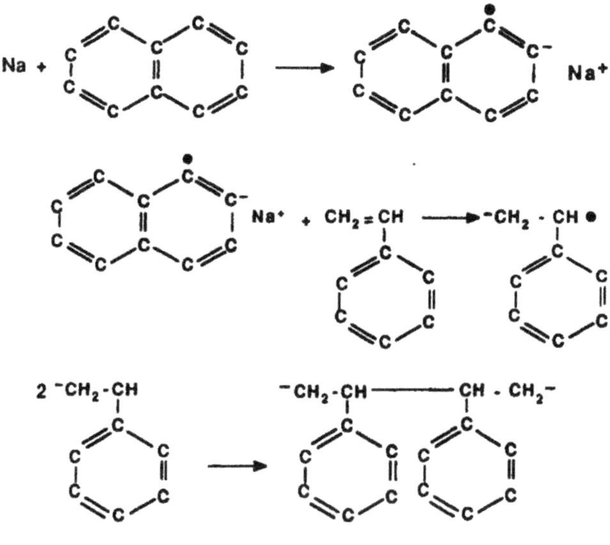

$$Na \ + \ (naphthalene) \ \longrightarrow \ (naphthalene \ radical \ anion) \ Na^+$$

$$(naphthalene \ radical \ anion) \ Na^+ \ + \ CH_2{=}CH(C_6H_5) \ \longrightarrow \ {}^-CH_2\text{-}CH(C_6H_5){\bullet}$$

$$2 \ {}^-CH_2\text{-}CH(C_6H_5) \ \longrightarrow \ {}^-CH_2\text{-}CH(C_6H_5)\text{-}CH(C_6H_5)\text{-}CH_2{}^-$$

The dianion is first reacted with styrene monomer and then with ethylene oxide, sequentially, to produce the triblock co-polymer:

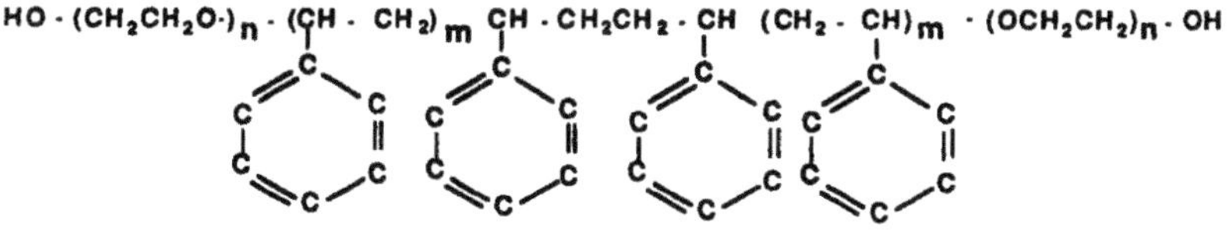

$$HO \cdot (CH_2CH_2O\cdot)_n \cdot (CH \cdot CH_2)_m \quad CH \cdot CH_2CH_2 \cdot CH \quad (CH_2 \cdot CH)_m \cdot (OCH_2CH_2)_n \cdot OH$$

Star-block copolymers (173) have been synthesized by in-dependently preparing living polystyryl using a 1-phenylethyl-potassium initiator and poly(ethylene oxide) with the same ini-tiator. A four-arm star-block is prepared by first reacting polystyrylpotassium with silicon tetrachloride at a polystyryl-potassium-to-silicon tetrachloride mole ratio of 2:1. That prod-uct is then sequentially reacted with living poly(ethylene oxide) again at a 2:1 mole ratio. A three-arm star can be prepared by a similar procedure with trichloromethylsilane in place of silicon tetrachloride.

$$
\begin{array}{c}
PEO \\
| \\
polystyryl{-}Si{-}polystyryl \\
| \\
PEO
\end{array}
\qquad
\begin{array}{c}
CH_3 \\
| \\
polystyryl{-}Si{-}polystyryl \\
| \\
POE
\end{array}
$$

Four-arm star   Three-arm star

Graft polystyrene/poly(ethylene oxide), poly(styrene-g-ethylene oxide) (174), has been prepared by making chloro-methylated polystyrene and condensing this product with living potassium-poly(ethylene oxide). The chloromethylpolystyrene can be prepared by chloromethylating polystyrene (175) or by copolymerization.

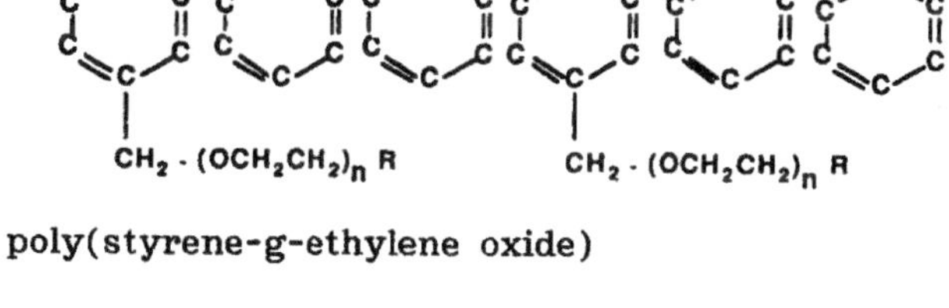

$$\cdot CH_2 \cdot CH \cdot CH_2 \cdot CH \cdot CH_2 \cdot CH \cdot CH_2 \cdot CH \cdot CH_2 \cdot CH \cdot CH_2 \cdot CH \cdot$$

$$CH_2 \cdot (OCH_2CH_2)_n R \qquad CH_2 \cdot (OCH_2CH_2)_n R$$

poly(styrene-g-ethylene oxide)

Ethylene oxide/styrene block copolymers have been further free-radical copolymerized with other ethylenically unsaturated compounds such as methyl methacrylate and methacrylic acid in benzene, tetrahydrofuran, and dimethylformamide (176). Correlations were made between reactivity ratio and solvent dielectric constant, as well as between solubility parameters of reaction solvent and growing polymer chains with marked effects apparent. Gel permeation chromatography of diblock and triblock copolymers based on polystyrene and poly(ethylene oxide) has revealed interesting molecular characteristics (177). Such block copolymers have an amphiphilic character. In aqueous solution, the polymers form spherical micells with a polystyrene core and a poly(ethylene oxide) outer sheath. The investigations used an aqueous-methanolic solution and were able to ascertain block copolymer structures and to estimate the impurities in the diblock copolymer.

Ethylene oxide has been grafted to a variety of substrates, such as collagen and gelatin (178) and silica (179). In the case of gelatin, grafting took place only on the free, primary amine groups of arginine, asparagine, and lysine with no monomer addition to the available $-OH$, $-COOH$, and $-CONH-$ groups. Other investigators (180) attempted grafting of acrylamide to poly(ethylene oxide) in dilute (0.1 percent) aqueous solution. When the aqueous solution was subjected to a hydrodynamic field that was imposed by stirring at 2100 rpm, grafting took place. However, in the absence of the field, no grafting occurred.

Sun and coworkers (181) used a different approach to prepare block copolymers. A polyoxyethylene with a terminal allyl group was reacted with acrylic acid to produce a graft copolymer polyoxyethylene-g-poly(acrylic acid).

Cherdon, Bodenbenner, and others have reported copolymerization of ethylene oxide and lactones (182, 183). With cationic initiation of valerolactone and ethylene oxide, a spiro intermediate was formed, leading to an alternating copolymer.

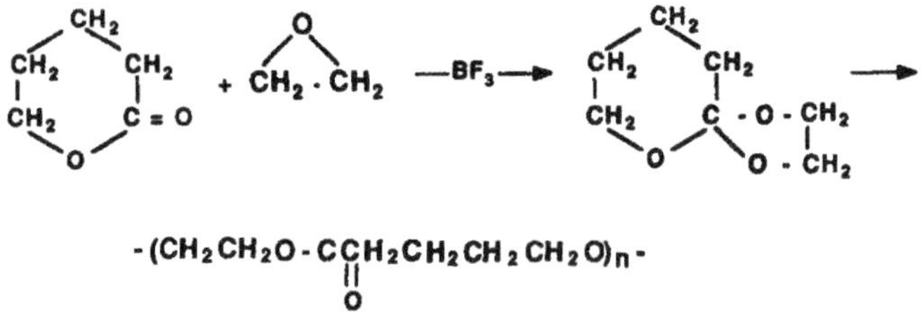

$$-(CH_2CH_2O-CCH_2CH_2CH_2CH_2O)_n-$$

   With dibutylzinc initiator and sequential polymerization, Bailey and France prepared block copolymers of ethylene oxide and epsilon-caprolactone and of propylene oxide and mixed methyl-epsilon-caprolactone (184). In these preparations, the poly-(ethylene oxide) segments were formed first, followed by epsilon-caprolactone addition.

   Linear block copolymers of the ABA type have been prepared using hydroxyl-terminated polyoxyethylenes as starter molecules and polymerizing epsilon-caprolactone onto each end of the molecules (185). Similar ABA block, polyester/poly(alkylene oxide)/polyester copolymers were synthesized by growing polyester blocks onto preformed polyoxyethylene by means of condensation polymerization (186). The polyester blocks included 1,6-hexanediol and 1,10-decanediol adipates and azelates.

   Branched or star-block copolymers were made (187) by first polymerizing ethylene oxide onto a starter using cationic initiation. These compounds were then coupled with the diglycidyl ether of bisphenol A to increase molecular weight and to provide pendant hydroxyl group functionality in the central portions of the alkylene oxide/glycidyl epoxide copolymer.

$$2\ HO-(CH_2CH_2-O)_n-H + (\underset{\underset{O}{\diagdown\diagup}}{CH_2-CH}-CH_2-O-C_6H_4)_2-C(CH_3)_2 \rightarrow$$

$$[HO-(CH_2CH_2-O)_n-CH_2-\underset{\underset{OH}{|}}{CH}-CH_2-O-C_6H_4]_2-C(CH_3)_2$$

These copolymers, tetra-hydroxyl functional in this example, were then reacted with lactone such as epsilon-caprolactone or epsilon-methyl-, gamma-methyl-, and gamma-t-butyl-epsilon-caprolactone to produce the branched or graft-block terpolymers exemplified below.

$$H\{[O(CH_2)_2\underset{\underset{C-(CH_2)_3}{|}}{CH}(CH_2)_2\overset{\overset{O}{\parallel}}{C}]_mO-(CH_2CH_2O)_n-CH_2-\underset{\underset{O[\underset{\underset{O}{\parallel}}{C}(CH_2)_2\underset{\underset{C-(CH_2)_3}{|}}{CH}(CH_2)_2O]_m}{|}}{CH}-CH_2-O-C_6H_4\}_2C(CH_3)_2$$

   The linear and branched block copolymers of alkylene oxides and caprolactones or polyesters are useful as disperse, acid, premetallized, and basic dye assistants; as antistatic agents;

and as antisoiling additives for polypropylene, cellulose acetate, and nylon fibers (186, 188, 189). Starlike macromers useful as engineering plastics have been prepared by copolymerizing lignin with propylene oxide (190). Multiblock copolymers from linear and branched alkyl acrylates containing five or more carbon atoms and ethylene oxide have been made (191). Various molecular weights and block lengths were prepared by first polymerizing ethylene oxide and then adding the specific acrylate and completing the polymerization process.

Polyester-polyether copolymers have been made by using an internal ester, such as 2,2-dimethyl-3-hydroxypropyl-2,2-dimethyl-3-hydroxypropionate (commonly known as Esterdiol 204), as a starter for the polymerization of ethylene and propylene oxide by means of cationic initiators (192).

$$HOCH_2-C(CH_3)_2CH_2OOC-C(CH_3)-CH_2OH \ + \ CH_2-HC-R \ \rightarrow$$
$$\underset{O}{\diagdown\diagup}$$

$$H(OCHRCH_2)_n-OCH_2-\overset{\overset{\displaystyle CH_3}{\displaystyle |}}{\underset{\underset{\displaystyle CH_3}{\displaystyle |}}{C}}CH_2OOC-\overset{\overset{\displaystyle CH_3}{\displaystyle |}}{\underset{\underset{\displaystyle CH_3}{\displaystyle |}}{C}}CH_2O-(CH_2CHRO)_nH$$

For these oligomers, it is preferred that n have a value of about 1 to 6. Modified block copolymers have been synthesized by coupling the above products with difunctional isocyanates (193) and with cycloaliphatic diepoxides (194). They were also used to synthesize radiation-curable diacrylates by means of a condensation reaction with acrylic acid (195). Other modifications of the oligomeric compounds include end capping with anhydrides to form acid-terminated oligomers (196) and reaction with polyols, anhydrides, and carboxylic acids to form low-viscosity, oil-free alkyds that are useful as coating intermediates (197).

Somewhat similar polyether copolymers have been prepared by polymerizing alkylene oxides onto various bisphenol compounds, such as 4,4'-dihydroxydiphenyldimethyl methane (198). The oligomeric compounds are useful as coating intermediates that can be crosslinked with aminoplasts.

Poly(N-acylimino)ethylene/poly(ethylene oxide)/poly(N-acylimino)ethylene ABA block copolymers have been prepared from polyoxyethylene dicarboxylic acid (199). ABA block copolymers have also been prepared from poly(1-proline) and amino-terminated polyoxyethylene (200).

REFERENCES

1. A. Wurtz, Ann. chim. phys. 69:330, 334 (1863).
2. H. Staudinger and H. Lohmann, Ann. chim. 500:41 (1933).
3. Y. Ishii and S. Sakai, in Ring Opening Polymerization (K. C. Frisch and S. L. Reegen, eds.), Marcel Dekker, New York, 1969, p. 1.
4. E. J. Vandenberg; T. Nakata; H. C. W. M. Buys, H. G. J. Overmars, and J. G. Noltes; O. Vogl, P. Loeffler, D. Branseleben, and J. Muggee; in Coordination Polymerization (C. C. Price and E. J. Vandenberg, eds.), Plenum Press, New York, 1983.
5. I. Kuntz, Transaction of New York Academy of Sciences, 33 (5):529 (1971).
6. F. E. Bailey, Jr., and J. V. Koleske, Poly(ethylene oxide), Academic Press, New York, 1976.
7. M. E. Pruitt and J. M. Baggett, U.S. Patent No. 2,706,181 (1955).
8. C. C. Price and M. Osgan, J. Amer. Chem. Soc. 78 (690):4787 (1956).
9. F. N. Hill, F. E. Bailey, Jr., and J. T. Fitzpatrick, Ind. Eng. Chem. 50:5 (1958).
10. F. E. Bailey, Jr., and H. G. France in Macromolecular Synthesis (N. G. Gaylord, ed.), John Wiley & Sons, Inc., Vol. 3, 1969, p. 77.
11. A. Eastham, in The Chemistry of Cationic Polymerization (P. H. Plesch, ed.), Macmillan, New York, 1963, p. 401.
12. P. Drefuss and M. P. Drefuss, in Chemical Kinetics (G. H. Bamford and C. F. H. Tipper, eds.), Vol. 15, Elsevier, Amsterdam, 1976.
13. R. A. Nelson and R. S. Jessup, J. Res., Natl. Bur. Stnds. 48:206 (1952).
14. R. E. Parker and N. S. Isaacs, Chem. Rev. 59:737 (1959).
15. G. Gee, W. C. E. Higginson, P. Levesley, and K. J. Taylor, J. Chem. Soc.:1338 (1959).
16. F. S. Dainton and K. J. Ivin, Quart. Rev. 12:61 (1958).
17. E. J. Vandenberg, J. Polymer Sci. 47:486 (1960).
18. E. J. Vandenberg, J. Polymer Sci., Part A-1, 7:525 (1969).
19. H. Meerwein, E. Battenberg, H. Gold, E. Pfell, and G. Willfang, J. Prakt. Chem. 154:83 (1939).
20. F. S. Dainton, R. E. Devlan, and P. A. Small, Trans. Fad. Soc. 76:467 (1954).

21. P. H. Plesch, The Chemistry of Cationic Polymerization, Pergammon Press, New York, 1963.

22. J. P. Kennedy and E. Marechal, Carbocationic Polymerization, John Wiley & Sons, New York, 1968.

23. S. Sakai, T. Sugiyama, and Y. Ishii, Kogyo Kagaku Zasshi 69:699 (1966).

24. G. T. Merrall, G. A. Latremouille, and A. M. Eastham, Can. J. Chem. 82:120 (1960); see also Reference 21, Chapter 10.

25. F. E. Bailey, Jr., unpublished work (1953).

26. H. Soler, V. Cadiz, and A. Serra, Angew. Makromol. Chem. 152:55 ((1987).

27. Y. Okamoto, Polymer Preprints 25 (1):264 (1984).

28. S. H. Yu, Polymer Preprints 25 (1):117 (1984).

29. S. Penczek, P. Kubisa, and K. Matyjaszewski, Advances in Polymer Science: Cationic Ring-Opening Polymerization 37, Springer Verlag, Berlin, New York (1980).

30. K. Kaeriyama, J. Poly. Sci.: Chem. Ed. 14:1547 (1976).

31. G. Gee, W. C. E. Higginson, and G. T. Merrall, J. Chem. Soc.:1345 (1959).

32. G. Gee. W. C. E. Higginson, K. Taylor, and M. W. Trenholme, J. Chem. Soc.:4298 (1961).

33. Y. Ishii, S. Sekiguchi, and A. Hayakawa, Kogyo Kagaku Zasshi 65:1041 (1962).

34. B. Wojtech, Makomol. Chem. 66:180 (1966).

35. Shigematsu, Y. Miura, and Y. Ishii, Kogyo Kagaku Zasshi 65:360 (1962).

36. S. Shigematsu, M. Suzuki, and Y. Ishii, Kogyo Kagaku Zasshi 64:153 (1961).

37. J. D. Ingram, D. D. Lawson, S. L. Manatt, N. S. Rapp, and J. P. Hardy, J. Macromol. Chem. 1:75 (1961).

38. L. E. St. Pierre and C. C. Price, J. Amer. Chem. Soc. 78:3432 (1956).

39. G. J. Pege, R. L. Harris, and J. S. MacKenzie, J. Amer. Chem. Soc. 81:3374 (1959).

40. D. M. Simmons and J. J. Verbanc, J. Polymer Sci. 44:303 (1960).

41. P. J. Flory, J. Amer. Chem. Soc. 62:1562 (1940).

42. M. Szwarc, Carbanions, Living Polymers and Electron-Transfer Processes, John Wiley & Sons, New York, 1968.

43. P. J. Flory, Principals of Polymer Chemistry, Cornell University Press, Ithaca, New York, 1953.

44. F. W. Billmeyer, Textbook of Polymer Science, Third Edition, Wiley-Interscience, New York, 1984.

45. B. Weibull and B. Nycander, Acta Chem. Scand. 8:847 (1954).
46. H. Shigematsu, Y. Miura, and Y. Ishii, Kogyo Kagaku Zasshi 63:360 (1962); Y. Ishii, S. Sekiguchi, and A. Hayakawa, Kogyo Kagaku 65:1041 (1962); S. Shigematsu, M. Suzuki, and Y. Ishii, Kogyo Kagaku Zasshi 64:1583 (1961).
47. J. Furukawa and T. Saegusa, Polymerization of Aldehydes and Oxides, Wiley-Interscience, New York, 1963.
48. M. Osgan and C. C. Price, J. Polymer Sci. 34:153 (1959).
49. D. G. Stewart and E. T. Barrow, U.S. Patent No. 2,870,099 (1959).
50. D. G. Stewart, U.S. Patent No. 2,870,110 (1959).
51. Y. Ishii, S. Sekiguchi, and M. Hattori, Kogyo Kagaku Zasshi 64:1497 (1961).
52. R. A. Miller and C. C. Price, J. Polymer Sci. 46:455 (1960).
53. J. Furukawa, T. Saegusa, and Y. Yokota, Pure Appl. Chem. 4:387 (1962).
54. J. Furukawa, T. Saegusa, T. Tsuruta, R. Sakata, G. Kakogawa, A. Kawasaki, and T. Harata, Kogyo Kagaku Zasshi 62:1269 (1959).
55. T. Kagiya, T. Shimizu, T. Sano, and K. Fukui, Kogyo Kagaku Zasshi 66:1148 (1963).
56. T. Kagiya, M. Hatta, T. Sano, T. Shimizu, and K. Fukui, Kogyo Kagaku Zasshi 66:1152 (1963).
57. K. Okazaki, Macromol. Chemie 43:84 (1961).
58. R. Sakata, T. Tsuruta, T. Saegusa, and J. Furukawa Makromol. Chemie 40:64 (1960).
59. J. Furukawa, T. Saegusa, T. Tsuruta, and G. Kakogawa, Makromol. Chemie 36:25 (1960); J. Polymer Sci. 36:541 (1959).
60. F. E. Bailey, Jr., U.S. Patent No. 2,914,419 (1959).
61. F. N. Hill, F. E. Bailey, Jr., and J. T. Fitzpatrick, U.S. Patent No. 2,987,489 (1961); Ind. Eng. Chem. 50:5 (1958).
62. F. N. Hill, F. E. Bailey, Jr., and J. T. Fitzpatrick, Belg. Patent No. 557,883 (1957).
63. F. N. Hill, F. E. Bailey, Jr., and J. T. Fitzpatrick, U.S. Patent No. 2,987,988 (1961).
64. F. N. Hill, F. E. Bailey, Jr., and J. T. Fitzpatrick, U.S. Patent No. 2,941,963 (1961).
65. F. N. Hill, F. E. Bailey, Jr., and J. T. Fitzpatrick, U.S. Patent No. 2,969,402 (1961).

66. F. N. Hill and J. T. Fitzpatrick, U.S. Patent No.
    2,866,761 (1961).
67. R. O. Colclough, G. Gee, W. C. E. Higginson, J. B.
    Jackson, and M. Litt, J. Polymer Sci. 34:171 (1959).
68. A. B. Borkovec, J. Org. Chem. 23:828 (1959).
69. E. C. Steiner, R. R. Pelletier, and R. O. Truck, J.
    Chem. Soc. 4678 (1964).
70. G. Gee, W. C. E. Higginson, and J. B. Jackson,
    Polymer 3:231 (1962).
71. E. J. Vandenberg, U.S. Patent No. 3,158,580 (1964).
72. T. Tsuruta, T. Hagiwara, and M. Ishimori, in Coordina-
    tion Polymerization (C. C. Price and E. J. Vandenberg,
    eds.), Plenum Press, New York, 1983.
73. S. G. Davies, Tetrahedron 42:5123 (1986); Tetrahedron
    Ltrs. 27:3787 (1986); particularly syntheses with propylene
    oxide using catalytic reagents as the complex of cyclopen-
    tadienyl-acetyl-iron carbonyl and triphenylphosphine.
74. D. A. Evans, Science 240:420 (1988).
75. R. Grubbs, Polymer Preprints 30 (1):17 Symposium,
    "Transition Metal Catalyzed Polymerizations: Mechanisms
    and Synthetic Utility," Dallas, Texas, 1989.
76. E. J. Vandenberg, J. Polymer Sci. B2:1085 (1964).
77. C. C. Price and R. Spector, J. Amer. Chem. Soc. 87:
    2069 (1965).
78. E. Stanley and M. Litt, J. Polymer Sci. 78:4787 (1956);
    J. Schafer, in Topics in 13C NMR Spectroscopy (G. C.
    Levy, ed.), Vol. 1, Wiley-Interscience, New York, 1974,
    p. 159.
79. Y. Kumata, N. Asada, G. M. Parker, and J. Furukawa,
    Makromol. Chemie 136:291 (1970).
80. F. E. Bailey, Jr., and K. D. Cavender, unpublished
    research, 1959.
81. J. Furukawa, T. Tsuruta, R. Sakata, T. Saegusa, and
    A. Kawasaki, Makromol. Chemie 32:90 (1959).
82. R. Sakata, T. Tsuruta, T. Saegusa, and J. Furukawa,
    Kogyo Kagaku Zasshi 63:1817 (1960).
83. V. Matsui, T. Hashimoto, T. Saegusa, and J. Furukawa,
    Kogyo Kagaku Zasshi 69:1375 (1966).
84. N. Kawabata, J. Furukawa, A. Kato, M. Nakaniwa, and
    A. Kawasaki, International Symposium on Macromolecular
    Chemistry, Tokyo, 1966.
85. M. Ishimori and T. Tsuruta, Makromol. Chemie 64:190
    (1963); Int. Symposium on Macromol. Chem., Kyoto, 1966.

86. J. Lal, J. Polymer Sci. Part A, 4:1163 (1966).
87. N. S. Chu and C. C. Price, J. Polymer Sci. Part A, 1: 1105 (1963).
88. E. J. Vandenberg, J. Polymer Sci. Part A, 7:525 (1969).
89. J. Furukawa and Y. Kumata, Makromol. Chemie 136:147 (1970).
90. M. Nakaniwa, K. Ozaki, and J. Furukawa, Makromol. Chemie 138:197 (1970).
91. M. Nakaniwa, I. Kaneoka, K. Ozaki, and J. Furukawa, Makromol. Chemie 138:209 (1970).
92. S. Inoue, T. Tsuruta, and N. Yoshida, Makromol. Chemie 79:34 (1964).
93. E. J. Vandenberg, Pure and Applied Chem. 48:295 (1976); D. A. Berta and J. Vandenberg, in Handbook of Elastomers: New Developments ((A. K. Bhowmish and H. L. Stephens, eds.), Marcel Dekker, New York, 1988, pp. 643–659.
94. F. E. Bailey, Jr., U.S. Patent No. 3,031,439 (1962).
95. E. J. Vandenberg, J. Polymer Sci. A1 7:525 (1969).
96. T. Tsuruta, J. Polymer Sci. Polymer Symposium 67:73 (1980).
97. H. Haubenstock, V. Panchalingan, and G. Odian, Makromol. Chemie 188:2789 (1987).
98. T. Aida, R. Mizuta, Y. Yoshida, and S. Inoue, Makromol. Chem. 182:1073 (1981).
99. Y. Yoo and J. E. McGrath, Polymer Preprints, Am. Chem. Soc. Polymer Chem. Div. 28 (2):360 (1987).
100. S. Inoue, J. Macromol. Sci.-Chem. A25 (5-7):571 (1988).
101. T. Aida, Y. Maekawa, S. Asano, and S. Inoue, Macromolecules 21 (5):1195 (1988).
102. S. Perry and H. Hibbert, J. Am. Chem. Soc. 62:2599 (1940).
103. R. Nomura, Y. Wada, and H. Haruo, J. Polymer Sci., Part A: Polymer Chem. 26 (2):627 (1988).
104. I. Kuntz, C. Cozewith, H. T. Oakley, G. Via, H. T. White, and Z. W. Wilchinsky, Macromol. 4:4 (1971).
105. F. E. Bailey, Jr., and H. G. France, J. Polymer Sci. 45:243 (1960).
106. T. Hagiwara, M. Takeda, H. Hamana, and T. Narita, Macromolecules 22:2026 (1989).
107. N. Yoshino, C. Suzuki, H. Kobnayashi, and T. Tsuruta, Makromol. Chem. 189:1903 (1988).
108. H. Kageyama, K. Miki, N. Tanaka, N. Kasai, M. Ishimori, T. Heki, and T. Tsuruta, Makromol. Chem., Rapid Commun. 3:947 (1982).

109. J. Milgrom, U.S. Patent No. 3,278,457 (1966), U.S. Patent No. 3,278,458 (1966), U.S. Patent No. 3,278,459 (1966), U.S. Patent No. 3,654,224 (1972); R. J. Herold, U.S. Patent No. 3,829,505 (1974).

110. A. Bencini, A. Bianchi, E. Garcia-Espana, M. Gusti, S. Mangano, M. Stephano, P. Orioli, and P. Paoletti, Inorg. Chem. 26:3902 (1987).

111. D. F. Mullica, W. O. Milligan, G. W. Beall, and W. L. Reeves, Acta Crystl. 17:3558 (1978).

112. L. Pauling and P. Pauling, Proc. Nat. Acad. Sci. 60:362 (1968).

113. H. E. Williams, Cyanogen Compounds, Edward Arnold and Co., London, 1948.

114. H. Fisher and G. Cuntz, Chem. Ztg. 26:872 (1902).

115. J. Hanford, U.S. Patent No. 2,402,137.

116. G. L. Goeke and F. J. Karol, U.S. Patent No. 4,193,892 (1980).

117. R. Lamot, Makromol. Chem. 189:45 (1988).

118. Y. Gnanou, P. Lutz, and P. Rempp, Makromol Chem. 189:2885 (1988).

119. A. Mouflou, J. G. Zilliox, G. Beinert, Ph. Chaumont, and J. Herz, in Biological Synthesis of Polymer Networks (O. Kramer, ed.), Elsevier Applied Science, London, 1988, pp. 483–493.

120. A. Stolarzewicz, Z. Grobelny, G. N. Arkhipovich, and K. S. Kazanskii, Macromol. Chem., Rapid Comm. 10:131 (1989).

121. F. T. Wall, J. Amer. Chem. Soc. 63:803 (1941).

122. F. R. Mayo and F. M. Lewis, J. Amer. Chem. Soc. 66:1594 (1944); T. Alfrey and G. Goldfinger, J. Chem. Phys. 12:205 (1944).

123. R. J. Gutter, J. Org. Chem. 26:2828 (1961); M. Haupstein and J. M. Lesser, J. Am. Chem. Soc. 78:676 (1956); D. B. Coffman, U.S. Patent No. 2,516,960 (1950); G. W. Stanton and C. E. Lowery, U.S. Patent No. 2,556,048 (1951).

124. T. Saegusa, T. Ueshima, and S. Tomita, Makromol. Chemie 107:131 (1967).

125. M. A. Bulgakova, K. I. Bolyschevskaya, and M. I. Siling, Poluch. Svoistva Primen. Plastmass Osn. Reaktsionnosposobn Oligomerov 39:143 (1978); CA 93:8633 (1980).

126. Z. Kalny and E. W. Kicko, Polimery (Warsaw) 24 (1):20 (1979); CA 91:21249 (1979).

127. PPG Industries, U.S. Patent No. 3,374,208 (1968).

128. W. Kuran and A. Nieslochowski, Polym. Bul. (Berlin) 2 (6):411 (1980).

129. W. Kuran and A. Nieslochowski, J. Macromol. Sci., Chem. A, 15 (8):1567 (1981).

130. R. D. Deanin and V. G. Dossi, Advan. Chem. Ser. 128: 176 (1973).

131. K. B. Wagener, C. Thompson, and S. Wanigatunga, Macromolecules 21:2668 (1988).

132. K. B. Wagener and S. Wanigatunga, Am. Chem. Soc. Symposium Ser. 364:153—164 (1988).

133. S. Wanigatunga and K. B. Wagener, Polymer Preprints, Am. Chem. Soc., Div. Polymer Chem. 29 (2):63 (1988).

134. S. Matsumura, N. Yoda, and S. Yoshikawa, Makromol. Chem., Rapid Comm. 10:63 (1989).

135. A. Hilt, J. Trivedi, and K. Hamann, Makromol. Chemie 66:177 (1965); A. Hilt and K. Hamman, Makromol. Chemie 92:55 (1966).

136. Air Products and Chemicals Inc., Bulletin, Poly(alkylene carbonates) - Typical Properties, Emmaus, PA, 1988, 5 p.

137. W. D. Willis, L. O. Amberg, A. E. Robinson, and E. J. Vandenberg, Rubber World 153:88 (1965).

138. L. J. Young, in Polymer Handbook (J. Brandup and E. H. Immergut, eds.), John Wiley & Sons, New York, 1974, p. 387.

139. R. G. Ismailov, S. M. Aliev, M. R. Bairamov, M. N. Magerramov, S. G. Aliev, and G. G. Ghadzhiev, Azerb. Khim. Zh. 4:60 (1970); CA 75: 98863 (1971).

140. O. G. Akperov, E. A. Dzhafarova, N. E. Bashirova, and Z. A. Ekhtibarova, Uch. Zap. Azerb. Un-t. Ser. Khim. 1:24 (1975); CA 85: 94734 (1976).

141. E. J. Vandenberg, J. Polym. Sci., A1, 10:329 (1972).

142. K. Kamio, M. Kwana, and S. Nakado, U.S. Patent No. 3,509,678 (1970).

143. G. P. Speranza, M. Cusurida, and R. L. Zimmerman, U.S. Patent No. 4,316,9991 (1982).

144. L. M. Landoll, U.S. Patent No. 4,304,902 (1981).

145. V. F. Jenkins, M. J. Beeken, and S. Pennington, U.S. Patent No. 3,795,701 (1974).

146. H. Kubota, T. Suda, M. Fukui, and K. Inoue, Japan Patent 73/34,400 (1973); CA 81: 78519 (1974).

147. R. D. Lundberg and F. D. DelGuidice, U.S. Patent No. 3,646,170 (1972).

148. F. E. Critchfield, J. E. Hyre, and E. C. Stout, U.S. Patent No. 3,689,531 (1972).

149. S. Sei, A. Ito, and Y. Kawai, Japan Patent 72/10,067 (1972); CA 77: 75814 (1972).

150. E. Baeder and L. Rohe, Ger. Offen. 1,955,848 (1971).

151. F. X. Mueller, Jr., and J. D. Brown, U.S. Patent No. 3,578,642 (1971).

152. R. R. Charpentier, J. C. Mileo, M. Osgan, B. Sillion, and G. De Gaudemaris, Fr. Addn. 96,334 (1972).

153. V. F. Jenkins and S. A. Lee, Brit. Patent 1,375,032 (1974).

154. V. F. Laporte and D. A. Doherty, Brit. Patent 1,376,331 (1974).

155. J. Luna de Prada, Ger. Offen. 2,231,785 (1971).

156. J. Brochet, Ger. Offen. 2,909,463 (1979).

157. J. Brochet, J. M. Huet, and P. Du Penhoat, Fifth Conf. Eur. Plast. Caoutch. 1:A13/1, Soc. Chim. Ind., Paris, France (1978).

158. H. L. Hsieh, O. G. Buck, and F. F. Naylor, U.S. Patent No. 3,867,353 (1975).

159. M. L. Kaufman, Ger. Offen. 2,643,701 (1977).

160. J. B. Niederl and V. Niederl, Micromethods of Quantitative Organic Analysis, John Wiley & Sons, New York (1946).

161. H. F. Rife and F. A. Roberts, U.S. Patent No. 2,448,664 (1948).

162. M. DeGroote, U.S. Patent No. 2,552,232 (1951).

163. E. C. Britton and P. S. Petrie, U.S. Patent No. 2,641,614 (1953).

164. J. T. Patton, U.S. Patent No. 2,786,080 (1957).

165. F. N. Hill, U.S. Patent No. 2,897,178 (1959).

166. Union Carbide Corporation, Technical Bulletin, TERGITOL[R] Surfactants, BASF-Wyandotte Corporation, Technical Bulletin, PLURONIC[R] Polyols.

167. Union Carbide Corporation, Technical Bulletin, NIAX[R] Polyols; BASF-Wyandotte Corporation, Technical Bulletin, PLURACOL[R] Polyols for Urethanes; Polyurethane Handbook (G. Oertel, ed.), Hanser-Verlag, Munich, 1985; F. E. Bailey, Jr., in Handbook of Polymeric Foams and Foam Technology (D. Klempner and K. C. Frisch, eds.), Carl Hanser Verlag, Munich, (in press).

168. F. Candau, F. Afshar-Taromi, and P. Remp, C. R. Acad. Sci. Paris C283:453 (1976).

169. J. Kelly, W. M. MacKenzie, D. C. Sherington, and G. Riess, Polymer 20:1048 (1979).

170. D. H. Richards and M. Szwarc, Trans. Fad. Soc. 56: 1644 (1959).

171. R. A. Quirk and J.-J. Ma, J. Polymer Sci., Part A: Polymer Chem. 26:2031 (1988).

172. T. Fang, S. Xu, and L. Yu, Yingyong Huaxue 188:2543 (1987).

173. Hongquan Xie and Jun Xia, Makromol. Chemie 188:2543 (1987).

174. M. H. George, M. A. Majid, J. A. Barrie, and I. Rezaian, Polymer 28:1217 (1987).

175. T. Alteres, D. P. Wyman, V. R. Allen, and K. Myerson, J. Polym. Sci. 13:4131 (1985).

176. J. Wang, C. Hu, and S. Ying, Gaofenzi Xuebao, 6:423 (1987); C.A. 109:23417 (1988).

177. I. V. Berlinova, N. G. Vladimirov, and I. M. Panayotov, Makromol. Chem., Rapid Comm. 10:163 (1989).

178. A. Gantar, A. Sebenik, M. Kavcic, U. Osredkar, and H. J. Harwood, Polymer 28:1403 (1987).

179. T. Tajouri, L. Facchini, A. P. Legrand, P. Balard, and E. Papirer, Chim. Phys. Phys.-Chem. Biol. 84 (2):243 (1987); C.A. 107: 59760 (1987).

180. B. P. Makogon, T. V. Stupnikova, and T. V. Vyshkina, Vysokomol. Soedin. Ser. B 30 (4):272 (1988).

181. F. Sun, M. Wang, H. Zhu, and S. Yang, Gaodeng, Xuexiao Huaxue Xuebao 8(7):658 (1987); C.A. 108: 38551 (1988).

182. H. Cherdon and H. Ohse, Makromol. Chemie 92:213 (1966).

183. K. Bodenbenner, Ann. 623:183 (1959).

184. F. E. Bailey, Jr., and H. G. France, U.S. Patent No. 3,312,753 (1967).

185. R. Perret and A. Skoulois, Comptes Rendu Acad. Sci. 268:230 (1969); Makromol. Chemie 156:143 (1972).

186. J. V. Koleske, R. M-J. Roberts, and F. D. Del Giudice, U.S. Patent No. 3,725,352 (1973).

187. J. V. Koleske, R. M-J. Roberts, and F. D. Del Giuidice, U.S. Patent No. 3,670,045 (1972).

188. J. V. Koleske, C. J. Whitworth, Jr., and R. D. Lundberg, U.S. Patent No. 3,781,381 (1973).

189. J. V. Koleske, R. M-J. Roberts, and F. D. Del Giuidice, U.S. Patent No. 3,825,620 (1974).

190. W. DeOliveira and W. G. Glasser, <u>J. Appl. Poly. Sci.</u> <u>37</u>:3119 (1989).

191. S. Hoering, R. D. Klodt, and H. Reuter, and J. Ulbricht, Ger. (East) DD262,866 (1988).

192. J. V. Koleske and R. J. Knopf, U.S. Patent No. 4,163,114 (1979).

193. O. W. Smith, J. V. Koleske, and R. J. Knopf, U.S. Patent No. 4,188,477 (1980).

194. O. W. Smith, J. V. Koleske, and R. J. Knopf, U.S. Patent No. 4,195,160 (1980).

195. R. J. Knopf and J. V. Koleske, U.S. Patent No. 4,163,113 (1979).

196. O. W. Smith, J. V. Koleske, and R. J. Knopf, U.S. Patent No. 4,171,423 (1979).

197. J. V. Koleske, U.S. Patent No. 4,297,476 (1981).

198. W. J. Blank, U.S. Patent No. 3,959,202 (1976).

199. M. Miyamoto, K. Naka, M. Tokumizu, and T. Saegusa, <u>Macromolecules</u> <u>22</u>:1604 (1989).

200. S. H. Jeon, S. M. Park, and T. Ree, <u>J. Poly. Sci.: Part A: Poly. Chem.</u> <u>27</u>:1721 (1989).

5

Chemistry of the Poly(alkylene oxide)s

I ANALYTICAL CHEMISTRY—END-GROUP DETERMINATIONS

ASTM D2849–69, Standard Methods of Testing Urethane Foam Polyol Raw Materials, describes a number of tests used for analyzing poly(alkylene oxide)s (1). These tests include metals analysis, acid and hydroxyl numbers, unsaturation values, water, suspended matter, specific gravity, viscosity, and color. Many of these tests are well known and will not be described. Those tests dealing with end groups—i.e., the functional groups that are used in the further reaction of poly(alkylene oxide)s—are discussed below. Detailed test procedures can be found in the cited reference and also can be obtained from poly(alkylene oxide) suppliers.

A. Hydroxyl Number

As discussed in Chapter 4, the hydroxyl number is a measure of the equivalent weight of free hydroxyl-containing compounds, and, if the functionality is known, the molecular weight can be determined. The equivalent or combining weight of a compound allows one to determine the amount of isocyanate that is needed to end-cap a polyol and prepare an isocyanate-terminated prepolymer,

$$H-(OCHR-CH_2)_n-X-(CH_2CHRO)_m-H + 2 OCN-Z-NCO \rightarrow$$

$$OCN-Z-\underset{\underset{HO}{|\;\|}}{N}COCHRCH_2-(OCHR-CH_2)_{n-1}-X-(CH_2CHRO)_{m-1}-CH_2CHRO\underset{\underset{OH}{\|\;|}}{C}N-Z-NCO$$

where X is a starter molecule as previously described and Z is a group such as diphenyl methane, to couple polyols and prepare a hydroxyl-terminated prepolymer,

$$2 H-(OCHR-CH_2)_n-X-(CH_2CHRO)_m-H + OCN-Z-NCO \longrightarrow$$

$$[H-(OCHR-CH_2)_n-X-(CH_2CHRO)_{m-1}CH_2CHRO\underset{\underset{OH}{\|\;|}}{C}N]_2-Z$$

and/or to prepare a thermoplastic or thermoset polyurethane from the polyol or prepolymers for use as foams, thermoplastic elastomers, cast elastomers, and so on.

There are a variety of methods for determining the hydroxyl numbers of polyols that involve capping the hydroxyl groups with a reactive compound and titrating a group that is formed. ASTM D2849—69 contains a description of the currently accepted methods used for polyether and polyester polyols (1). This method involves reaction of the hydroxyl groups with phthalic anhydride or acetic anhydride and titrating the carboxyl groups formed with a base. Wellons, Cary, and Elder (2) describe certain shortcomings of the method, such as a 2-hour reaction time, and suggest a far more rapid method, with a 15-minute reaction time, involving the use of imidazole as a catalyst and pyridine as a solvent/proton scavenger. Results obtained with alcohols, poly(alkylene oxide)s, and nonionic surfactants are presented, and the authors state that their rapid method is statistically equivalent to the ASTM method.

B. Acid Number

Acidity can cause retardation of reactions with isocyanates, such as polyols undergo in the preparation of polyurethanes. In addition, if such acidity is significant, it can affect the calculated

equivalent weight and lead one to add incorrect amounts of reactants such as isocyanates, which can cause property alteration. For this reason, the hydroxyl number should be cor-
rected for acidity by adding the acid number to the hydroxyl
number before determination of the equivalent weight. A similar correction for alkalinity, if present in the poly(alkylene oxide), should also be made.

   ASTM D2849 also describes a method for determining the
acid number and correcting for acidity or alkalinity in polyether
polyols. The acid number is defined as the milligrams of potassium hydroxide that are needed to titrate acidic constituents
present in 1 gram of the poly(alkylene oxide). A suitably sized
sample is dissolved in a solvent containing a phenolphthalein
indicator and titrated with potassium hydroxide.

C. Unsaturation

In Chapter 4, mechanisms by which carbon-to-carbon unsaturation can be introduced onto the ends of poly(alkylene oxide)s
were described. Since these end groups can have a deleterious
effect on the performance characteristics of polyurethane foams
or elastomers, as well as on other derivatives, the degree of
unsaturation is a specification often required by manufacturers
or a tool for understanding lot-to-lot variation in products produced.

   ASTM D2849 describes a method in which the poly(alkylene
oxide)s are reacted with mercuric acetate and methanol in a
methanolic solution. The reaction produces acetoxymercuric-
methoxy compounds and acetic acid. The acetic acid produced,
which is directly proportional to the amount of unsaturation, is
determined by titration with alcoholic potassium hydroxide in
the presence of a phenolphthalein indicator. As is the case
with other acidimetric titrations, a suitable correction must be
made if the starting sample is not neutral to the indicator.

II. HYDROXYL-GROUP REACTIONS

Most poly(alkylene oxide)s are terminated with hydroxyl groups
and are used as such unless modified in a subsequent reaction,
as with the amine-terminated products described in the next
section. Thus, reactions of the hydroxyl group are important
to the chemistry of the poly(alkylene oxide)s.

A. Ester Formation (3, 4)

An important hydrophobe for nonionic surfactants is obtained from fatty acid moieties. Of particular importance are poly(ethylene oxide) esters of the C_{12} to C_{18} fatty acids. Such compounds can be formed by the direct esterification of polyoxyethylene with a fatty acid via a condensation reaction.

$$HO-(CH_2CH_2O)_n-H + RCOOH \rightleftharpoons RCOO-(CH_2CH_2O)_n-H + H_2O$$

$$\begin{array}{cc} C_{12}-C_{18} & \text{polyoxyethylene} \\ \text{fatty acid} & \text{ester} \end{array}$$

A mixture of ester is obtained, and the ratio of monoester to diester is controlled by ratios of the compounds charged to the reactor. Excess polyoxyethylene is used to maximize monoester production (5), and excess fatty acid is used to maximize diester formation (6). Because of the existing equilibrium, it is important that water be removed with an azeotroping agent such as toluene, xylene, etc., and/or by use of an inert-gas sparge to carry off water as it is formed to force the equilibrium toward the desired product. Catalysts such as sulfuric acid (7), benzene sulfonic acid, and other aromatic sulfonic acids (5, 8, 9), as well as cationic ion-exchange resins such as polystyrene-sulfonic acids (5, 9), are used. The latter compounds have the advantage of easy removal from batch reactions and of use in a fixed bed for continuous processes. Metals such as tin, iron, and zinc, as well as their salts in powdered form, have been used as catalysts (10, 11). Catalysts can improve the yield of monoester. Of course, use of a monohydroxyl-functional polyoxyethylene, such as that from methanol-started ethylene oxide polymers (methoxy-polyoxyethylene), can be esterified with fatty acids to yield effectively all monoester.

The other commercial method for preparing polyoxyethylene esters does not involve reaction of the hydroxyl group of a preformed polyoxyethylene molecule. Rather, ethylene oxide is reacted with the fatty acid in the presence of an alkali catalyst such as an alkaline hydroxide, alcoholate, or metal (8, 12).

$$RCOOH + KOH \longrightarrow RCOO^-K^+ + H_2O$$

$$\begin{array}{ccc} \text{fatty acid} & \text{base} & \text{alcoholate} \end{array}$$

$$RCOO^-K^+ + \underset{\underset{O}{\diagdown\diagup}}{CH_2-CH_2} \longrightarrow RCOOCH_2CH_2O^-K^+$$

The alcoholate ions can then rapidly add ethylene oxide to produce the desired ester, with termination finally taking place by
oxide exhaustion or neutralization with a molecule of fatty acid.

$$RCOOCH_2CH_2O^-K^+ + RCOOH \rightarrow RCOOCH_2CH_2OH + RCOO^-K^+$$

Transesterification (12, 13) plays an important role in the preparation of these compounds, and mono- and diester as well as
polyoxyethylene glycols form.

Polyoxyethylene ester surfactants are used in both industrial and household formulations. Specific end-use product areas
include textile scouring agents, softening agents, dye assistants, and lubricants; emulsions and wettable powders used in
agricultural pest control; and home floor, rug, and wall clean-
ers, as well as laundering and dishwashing compositions. Specific formulations for such products can be found in the literature (3).

Chen and Liu (14) coupled low-molecular-weight (number
average values of 106 to 1000) polyoxyethylenes with 5-sulfo-
isophthalic acid dimethyl ester sodium salt by condensing methanol from the molecules and forming the corresponding hydroxyl-
terminated, ester-linked polyethers. These compounds were
then reacted with phthalic anhydride by means of addition and
condensation reactions to form water-soluble, polyester-linked
poly(ethylene oxide)s with molecular weights between 6000 and
9000. The products had excellent surface-active properties, as
well as excellent dispersant properties for disperse dyestuffs.

B. Polyurethanes and Associated Reactions

Isocyanates react with hydroxyl groups in a rearrangement reaction to form a urethane group. In the simple case, this can
be described by an alcohol reacting with an isocyanate, as indicated below.

$$R-NCO \quad + \quad R'-OH \quad \rightarrow \quad R-\overset{\displaystyle H}{\underset{\displaystyle |}{N}}-\overset{\displaystyle O}{\underset{\displaystyle \|}{C}}-O-R'$$

| isocyanate | alcohol | R and R' linked by a urethane group |

This reaction will take place at room temperature and below,
but in most commercial manufacturing processes involving reac-

tions between these compounds, thermal energy and a catalyst such as dibutyltindilaurate, stannous octanoate, zinc octanoate, or similar compound is used to catalyze the reaction. To make polyurethanes, which are widely used in industry as thermoplastic elastomers, cast thermoset elastomers, moisture-cure coatings, and foams for a variety of end uses that are described in a later chapter, multifunctional polyols and multifunctional isocyanates and short chain diols such as 1,4-butanediol are reacted together in various ratios. The polyols are usually termed the soft or flexible segments, and the reaction product of the isocyanates and short chain diols are termed the hard segment. The former segments impart flexibility and/or toughness to a polyurethane, and the latter segments act as pseudo-crosslinks in thermoplastic elastomers and also as stiffening and strengthening factors in thermoset elastomers and foams.

Preparation of a thermoplastic polyurethane can be exemplified by the following reaction, wherein a difunctional propylene oxide polyol, 4,4-diphenylmethane diisocyanate, and 1,4-butanediol are used; these reactants are designated as $[H-(PrO)_x-O]_2-R$, OCN$-$MD$-$NCO, and HO$-$Bu$-$OH, respectively.

$$[H-(PrO)_x-O]_2-R \quad + \quad 2 \ OCN-MD-NCO \ + \quad HO-Bu-OH \longrightarrow$$

$$-O-Bu-O\overset{O}{\overset{\|}{C}}-\overset{H}{\overset{|}{N}}-MD-\overset{H}{\overset{|}{N}}-\overset{O}{\overset{\|}{C}}-O\overset{H}{\underset{\underset{CH_3}{|}}{\overset{|}{C}}}CH_2-(PrO)_{x-1}ORO(PrO)_{x-1}-CH_2\overset{H}{\underset{\underset{CH_3}{|}}{\overset{|}{C}}}O-\overset{O}{\overset{\|}{C}}-\overset{H}{\overset{|}{N}}MD-\overset{H}{\overset{|}{N}}-\overset{O}{\overset{\|}{C}}-$$

This structure, on the average, continuously repeats until a high-molecular-weight (50,000 to 100,000 or more) elastomer is formed. A ratio of polyol/isocyanate/short chain diol of 1/2/1 was used in this example. Thus, the number of hydroxyl groups was equal to the number of isocyanate groups available for reaction. When this equality is maintained, a thermoplastic product results. If the amount of isocyanate is increased to greatly above the equality, a crosslinked thermoset product will result. If the amount of polyol or short chain diol is increased too much above the equality, molecular weight will suffer and a poor, weak product will result (15). Also, note that various ratios of the components can be used, and these ratios can be generalized as 1/x/x$-$1 where x is the number of moles of diisocyanate and of diol used (16). As x increases, the amount of hard segment and product hardness increases.

In the preparation of polyurethanes, isocyanates can be consumed in reactions other than with hydroxyl groups. If water (moisture) is present, it will react with an isocyanate to form carbamic acid, which is an unstable compound and slowly decomposes to an amine and carbon dioxide.

$$RNCO + H_2O \longrightarrow [RNHCO_2H] \longrightarrow RNH_2 + CO_2$$

The amine that is formed will rapidly react with another isocyanate group to produce a urea linkage and a substituted urea compound.

$$RNCO + RNH_2 \longrightarrow RNHCONHR$$

These are the reactions that take place in the polymerization or cure of moisture-curing polyurethane coating systems. Although of importance for such coatings, notice that two molecules of isocyanate are consumed for each molecule of water present and, because of its low molecular weight, even small amounts of water use considerable amounts of isocyanate and alter the final properties of the polyurethane being made. The reactions can also be of importance in foam manufacture, where the carbon dioxide can act as a blowing agent.

Isocyanate can also react with a urethane group to form an allophanate.

$$\underset{\text{urethane}}{RNCO + RNHCOOR} \longrightarrow \underset{\text{allophanate}}{R-N(CONHR)-COOR}$$

Reactions such as these occur at a slow rate at temperatures of less than 120°C to 140°C in uncatalyzed systems.

Isocyanates are more reactive with urea groups than with urethane groups and form biurets at a moderate rate at temperatures of about 100°C or higher.

$$\underset{\text{urea}}{RNCO + RNHCONHR} \longrightarrow \underset{\text{biuret}}{R-N(CONHR)-CONHR}$$

The formation of allophanates and biurets are important in that they will lead to branching in the polymeric network.

Isocyanates will also react with carboxylic acids to form an unstable mixed anhydride that decomposes to form an amide and carbon dioxide.

$$RNCO + R'COOH \longrightarrow [RNHCOOCOR'] \longrightarrow RNHCOR' + CO_2$$

mixed anhydride   amide

These reactions are usually undesirable, since they lead to a decrease in the amount of isocyanate available for reaction with hydroxyl groups and produce carbon dioxide, which can cause bubbles in the polyurethane if it is formed late in the polymer's formation.

C. Reaction with Halogens and Other Compounds

Much of the information in this section deals with block copolymers. Readers are also directed to the sections in Chapter 4 that deal with a variety of block copolymers.

Craven and coworkers (17) prepared ethylene oxide/oxymethylene block copolymers by means of a modified Williamson ether synthesis (18). Low-molecular-weight (~200) polyoxyethylenes were reacted with dibromomethane in the presence of potassium hydroxide in chlorobenzene solvent.

$$H(OCH_2CH_2)_4OH \; + \; Br-CH_2-Br \; \xrightarrow{\quad KOH, \; chlorobenzene \quad}$$

$$[OCH_2-(OCH_2CH_2)_4]_x + H-(OCH_2CH_2)_4-[OCH_2-(OCH_2CH_2)_4]_y-OH$$

$$+ \; HBr$$

Gel permeation chromatography and nuclear magnetic resonance studies indicated that both ring structures—with x varying from 1 to 15, indicating that rings with from 14 to more than 200 ring atoms were synthesized—and linear polymer—with a broad molecular weight distribution and molecular weights up to a million—were formed.

Polyoxyethylene has also been reacted successfully with a

difunctional, acid chloride-terminated prepolymer of poly(m-phenylene isophthalamide) (19). Other investigators prepared block copolymers from oligomeric poly(ethylene glycol-terephthalate) and polyoxyethylene and found that from the melt, the blocks crystallized independently and formed the crystal habit peculiar to those of the individual homopolymers (20). Wodka and Danielewicz (21) prepared block copolymers of polyoxyethylene and polyacrylonitrile from polyoxyethylene that contained azo or xanathate groups. It is apparent that fibers structurally designed from such block copolymers of these two polymers could have excellent antistatic properties, as well as improved and perhaps altered dye-acceptability characteristics.

Polyimides containing polyoxyethylene units as flexibilizers were prepared (22) by reacting 1-fluoro-4-nitrobenzene with ethylene oxide oligomers (n = 1 to 4 below) and then converting the nitro compound by means of hydrogenation into a diamine and these into the polyimides by reaction with 2,2-bis(3,4-dicarboxyphenyl)hexafluoropropane dianhydride.

$$F-C_6H_4-NO_2 \quad + \quad HO-(CH_2CH_2O)_n-H \quad \longrightarrow$$

$$O_2N-C_6H_4-O-(CH_2CH_2O)_n-C_6H_4-NO_2 \quad \xrightarrow{\text{hydrogen, Pd}}$$

$$H_2N-C_6H_4-O-(CH_2CH_2O)_n-C_6H_4-NH_2$$

The final compounds had improved flexibility, T_g (glass transition temperature) decreased from ~250°C to 140°C, but thermal stability (5 percent weight loss) decreased from 460°C to 385°C in air and from 475°C to 405°C in argon as n increased from 1 to 4.

An important reaction for epichlorohydrin is the reaction between it and bisphenol A, 4,4'-isopropylidinediphenol, to produce the highly important epoxides that are usually referred to as the diglycidyl ethers of bisphenol A to distinguish them as a class from the cycloaliphatic epoxides, Novolac epoxides, and so on.

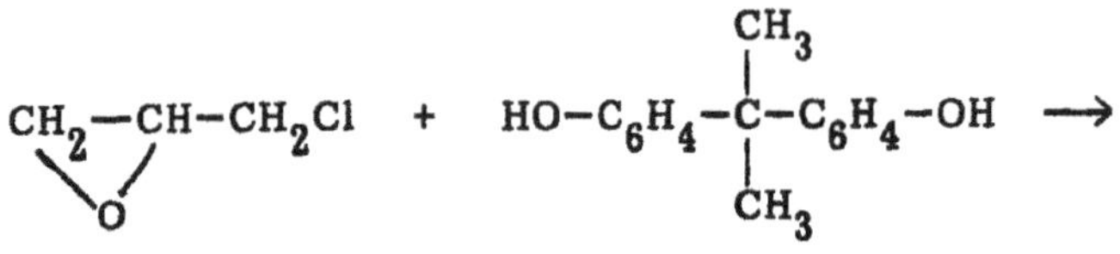

$$CH_2\!-\!CH\!-\!CH_2Cl \;+\; HO\!-\!C_6H_4\!-\!\overset{\overset{\displaystyle CH_3}{|}}{\underset{\underset{\displaystyle CH_3}{|}}{C}}\!-\!C_6H_4\!-\!OH \;\longrightarrow$$

epichlorohydrin   bisphenol A

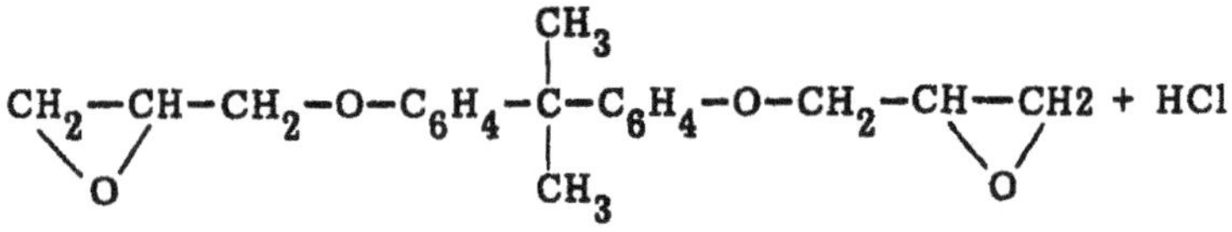

$$CH_2\!-\!CH\!-\!CH_2\!-\!O\!-\!C_6H_4\!-\!\overset{\overset{\displaystyle CH_3}{|}}{\underset{\underset{\displaystyle CH_3}{|}}{C}}\!-\!C_6H_4\!-\!O\!-\!CH_2\!-\!CH\!-\!CH2 \;+\; HCl$$

diglycidyl ether of bisphenol A

The hydrochloric acid formed in this reaction is neutralized
with sodium hydroxide. Although the above reaction scheme
suggests that 2 moles of epichlorohydrin are reacted with 1
mole of bisphenol A to obtain the desired product, in practice
it is found that a significant excess of epichlorohydrin is needed
to minimize reaction of the epoxide groups with hydroxyl groups
when the indicated, lowest member of the series is desired (23,
24). Manipulation of the ratio of epichlorohydrin to bisphenol
A and to caustic will yield a variety of products that have the
following generalized structure and that form a family of important commercial products.

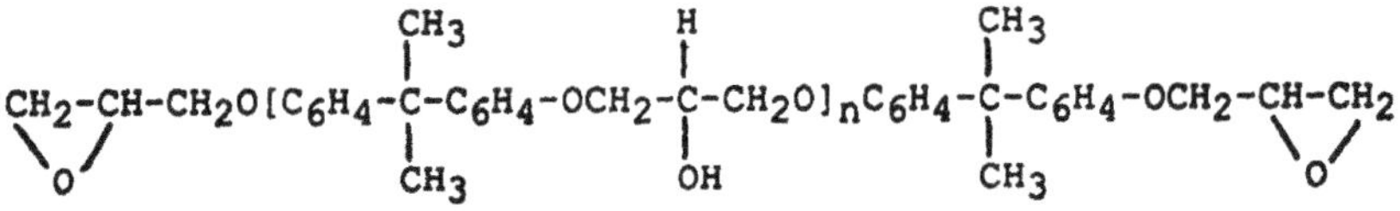

$$CH_2\text{-}CH\text{-}CH_2O\,[\,C_6H_4\text{-}\overset{\overset{\displaystyle CH_3}{|}}{\underset{\underset{\displaystyle CH_3}{|}}{C}}\text{-}C_6H_4\text{-}OCH_2\text{-}\overset{\overset{\displaystyle H}{|}}{\underset{\underset{\displaystyle OH}{|}}{C}}\text{-}CH_2O\,]_n\,C_6H_4\text{-}\overset{\overset{\displaystyle CH_3}{|}}{\underset{\underset{\displaystyle CH_3}{|}}{C}}\text{-}C_6H_4\text{-}OCH_2\text{-}CH\text{-}CH_2$$

Brominated and chlorinated bisphenol A are also epoxidized by
reaction with epichlorohydrin to make epoxides that are used
in flame-retardant coatings and shaped articles. A good example
of a shaped article is FR-4 board, which is a fiberglass/epoxide
composite that is the basis of many printed circuit boards.

    The reactivity of the pendant chloromethyl side group of
polyepichlorohydrin and copolymers containing epichlorohydrin
units has been used to make a variety of interesting functional
polymers. In aprotic solvents such as dimethyl sulfoxide, dimethylformamide, etc., these side groups undergo substitution
reactions with nucleophilic reactants such as amines (25) and

potassium cinnamate (26), and with potassium carbazole (27), and undergo elimination reactions with potassium t-butoxide (28). Modification of polyepichlorohydrin to form polymers with a high glass transition temperature (as high as about 60°C) and containing the vinyloxy moiety in the side chain, as well as pendant 2-thiobenzothiazole or phthalyl groups, has been made (29). Certain of these polymers are useful as ultraviolet light-activated, negative-type photoresists when 2,6-di-(4'-azobenzal)-4-methylcyclohexanone is used as a photosensitizer. Other polymers from this study are useful as ultraviolet light-activated, positive-type photoresists when cationic initiators such as 2,5-dibutoxy-4-morpholinobenzene diazonium hexafluoroantimonate are used.

Novel ABC block copolymers that contained a hydrophobic polymer, A, a hydrophilic spacing polymer B, and a bioactive agent, C, were prepared by Vulic and coworkers (30). The copolymer was prepared from monoamino-polystyrene, diamino-polyoxyethylene, and the biologically active, anticlotting agent heparin by means of coupling reactions using diisocyanates and carbodiimide activation of a carboxyl group in heparin. In the following descriptive equations, —U— designates a urea linkage.

$$P(Sty)-NH_2 \ + \ OCN-R-NCO \ \longrightarrow \ P(Sty)-U-R-NCO$$

$$P(Sty)-U-R-NCO \ + \ H_2N(EO)_nNH_2 \ \rightarrow \ P(Sty)-U-R-U-(EO)_nNH_2$$

$$P(Sty)-U-R-U-(EO)_nNH_2 \ + \ HEP-COOH \ \xrightarrow{\text{carbodiimide}}$$

$$P(Sty)-U-R-U-(EO)_nNHCO-HEP$$

anchorable or fabricable
polymer that contains a
hydrophilically spaced,
biologically active compound

The goal was to prepare an anchorable (polystyrene) product that, when exposed to an aqueous, saline environment, would result in exposure and swelling or other freedom of the hydrophilic spacer that would, in turn, allow the biologically active compound, heparin, to inhibit clotting and/or that could be used for fabrication of blood-contacting devices for the same purpose. These potential end uses were to be tested at a later date.

Cyano-terminated poly(propylene oxide)s have been prepared by reacting propylene oxide polyols with acrylonitrile in
the presence of water and sodium hydroxide (31). The ratio of
water to caustic had an effect on the degree of capping. New
approaches to the synthesis of functional poly(ethylene oxide)
by functionalization during the anionic initiation or deactivation
of a living polymer have been described (32). Dinitrophenyl
derivatives of 3×10^3 to 10^5 molecular weight, which function
as synthetic immunogenes, were prepared. Vinyl ether termination was also achieved, and these macromonomers were useful
for grafting poly(ethylene oxide) onto surfaces.

Block copolymers of 1,2-butylene oxide and ethylene oxide
have been prepared and evaluated for emulsion stability (33).
The effect of copolymer composition and molecular weight over
the range of 500 to 1000 on emulsifying power was determined.

From one to four units of ethylene oxide have been used to
space a group with liquid crystalline properties, methoxybi-
phenyl, different distances from a methacrylate group (34).
Hydrodynamic and thermodynamic properties were calculated.

D. Graft Poly(alkylene oxide)s

When the discussion in Chapter 4 is considered, it is readily
apparent that the majority of molecules that contain hydroxyl
functionality can be used as starters for the polymerization of
alkylene oxides. These starters can be low molecular weight,
oligomeric, or polymeric in nature. Many of the block copolymers described above involved addition of alkylene oxides or
other heterocyclic compounds such as the lactones to terminal
hydroxyl groups of preformed polymers. If the hydroxyl groups
of a starter are dispersed along a macromolecular chain and ethylene oxide is added to them, graft copolymers result. For example, if an alkylene oxide were added to a styrene-allyl alcohol
copolymer, the following graft copolymer would result.

$$— (CH_2-\underset{\underset{C_6H_5}{|}}{CH}-CH_2-\underset{\underset{CH_2OH}{|}}{CH}-CH_2-\underset{\underset{C_6H_5}{|}}{CH}-\underset{\underset{CH_2OH}{|}}{CH_2CH}) — \quad + \quad CH_2\overset{\displaystyle\diagdown}{\underset{O}{}}CHR \longrightarrow$$

$$— (CH_2-\underset{\underset{\underset{H-(OCHRCH_2)_a-O}{|}}{CH_2}}{CH}-CH_2-\underset{\underset{C_6H_5}{|}}{CH}-CH_2-\underset{\underset{O-(CH_2CHRO)_b-H}{CH_2}}{CH}-\underset{\underset{C_6H_5}{|}}{CH_2-CH}) —$$

A classic example of a graft copolymer is hydroxyethyl-cellulose (35, 36). The graft copolymer was first disclosed by Dreyfus (37) in 1921 and is presently an important commercial product used in various industries, such as architectural coatings, textile, paper, gas and oil drilling, and adhesives. The copolymer is prepared by a slurry process (38) or a vacuum process (39). In either case, cellulose is made more accessible by a swelling treatment in aqueous caustic and then hydroxyalkylating with either ethylene chlorohydrin or ethylene oxide, with the latter preferred in modern processes. As the amount of ethylene oxide is increased, water sensitivity is increased until solubility in dilute alkali is achieved when more than about one mole of ethylene oxide is added per mole of anhydroglucose units. When the molar substitution is greater than 1.6, the graft copolymer becomes ethanol soluble (40). The characteristics of hydroxyethylcellulose in aqueous solution have been the subject of many investigations by Brown and coworkers, as well as others (40−45). It has also been studied in a highly cross-linked, aqueous gel form (46). Hydroxyethylcellulose that was water soluble and alcohol/methylene chloride soluble, depending on degree of substitution, has been made (47, 48). Hydroxypropyl-methyl-cellulose that had a greater solubility in organic solvents and improved thermoformability over that of hydroxyethyl-ethyl-cellulose has been described by Schick (49).

Chitin is another naturally occurring polymer that has been ethoxylated. Chitin differs from cellulose in that the hydroxyl group on Carbon 2 of the anhydroglucose units has been replaced with an acetamide group.

Cellulose   Chitin

Chitin was first swollen by treatment with alkali and then alkoxylated with ethylene oxide under homogeneous (50, 51) or heterogeneous (52) conditions to produce water-soluble polymers.

Oligomeric ethylene oxide terminated with 1,2,4-triazoline-3,5-dione has been grafted onto polybutadiene at room temperature in a mixture of tetrahydrofuran and ethylene chloride (53). The product had a broad molecular weight distribution, and energy absorption maxima related to the glass transition temperature of each polymer were apparent.

III.   AMINE-TERMINATED POLY(ALKYLENE OXIDE)S

Amine-terminated poly(alkylene oxide)s have been made and used in industry for over half a century.   These compounds can be prepared in various ways.   Lee and Winfrey (54) describe a process in which poly(propylene oxide)s are converted into diamines by reaction with anhydrous ammonia, hydrogen and anhydrous ammonia, or ammonium hydroxide in the presence of a Raney nickel catalyst.   In the following

$$\{H-[O-CH(CH_3)CH_2]_x-O\}_2-R \ + \ NH_3 \ + \ H_2 \ \xrightarrow{\text{Raney nickel}}$$

$$\{H_2NCH(CH_3)CH_2-[O-CH(CH_3)CH_2]_{x-1}-O\}_2-R$$

reaction scheme, R is the residue of the starter molecule.   End uses described include curing agents for diglycidyl ethers, brake-fluid additives because the amines did not corrode metals or soften rubbers, and reactants with polycarboxylic acids and with polyisocyantates to form solid polyamides and solid polyureas, respectively.

Yeakey (55) described an improved method for preparation of polyoxyalkylene-polyamines that involved a reductive amination of polyoxyalkylene-polyols in the presence of a nickel-copper-chromium catalyst.   When the poly(alkylene oxides)s in combination with ammonia and hydrogen were contacted with the catalyst at elevated temperatures (200–250°C) under pressure (2000–4000 psi 13.78 to 27.56 MPa), yields of the corresponding diamine, which was predominantly primary in nature, improved over that of other known processes.   In addition, the process provided for the amination of higher molecular weight polyols than was possible before.   Primary amines were the main reaction product when polyols with terminal, secondary hydroxyl groups were used.   For this reason, it was preferred that poly-

(alkylene oxide) chains were terminated with at least one mole
of propylene oxide.

An early end use for amine-terminated poly(alkylene oxide)s
involved the use of tri- and tetraethylene glycol diamine as
compounds that improved the dyeability of fiber-forming polyamides (56). The use of the same primary amines, as well as
diethylene glycol bis(propylamine) and a poly(propylene glycol)-
diamine as coreactants (curing agents) with the diglycidyl ether
of 2,2-bis(p-hydroxyphenyl)propane, has been described (57).
The primary amines improved reactivity and flexibility of the
cured epoxide systems. Bishop and coworkers (58) describe
the utility of polyoxyalkylene diamines with molecular weights of
up to about 2500 as intermediates in the preparation of acrylates
that are used as ultraviolet light-curable buffer coatings and
top coatings for optical fibers.   Other end uses for the ami-
nated compounds include selective recovery of carbon dioxide
from acid gas streams (59), use as components in cationic elec-
trodeposition coating compositions for rough steel (60), use as
low-temperature flexibilizers in copolyetheramide block copolymers (61), use as fabric antistatic agents that are introduced
during a laundering cycle (62), and use as tensile-elongation
improvers for polyetherimide esters (63).

Polyoxyethylene molecules with tertiary amino end groups
were prepared by methylation of amino-terminated glycols (64).
These compounds could be transformed into quaternary ammonium groups with $(CH_3O)_2SO_2$ and then into chlorides with a
basic anion exchanger. Block copolymers have been prepared
from aromatic polyamides and polyoxyethylene (65).

IV.   POLYMER POLYOLS, FREE-RADICAL
GRAFTED POLY(ALKYLENE OXIDE)S

In 1966, a new class of polyols that were highly useful in en-
hancing the modulus of polyurethane foams and elastomers,
while maintaining other desirable properties, was introduced to
the marketplace (66, 67). These polyols had the unique feature
of containing in situ, free-radical polymerized vinyl polymer
particles that were grafted to the polyol. The final product,
which was termed a polymer polyol, is a conventional or an ethylene oxide-capped poly(propylene oxide) polyol that contains a
stable dispersion of the vinyl polymer that acts as a reinforcing
filler. When monomers such as acrylonitrile and styrene/acry-

lonitrile mixtures are used, property enhancements are obtained
in subsequently manufactured products.

   Although various free-radical sources can be used to prepare polymer polyols, improved results in later manufacture
were obtained when azobis-isobutyronitrile (AIBN) was used
(66, 67). The formation of polymer polyols can be depicted in
the following simplistic manner.

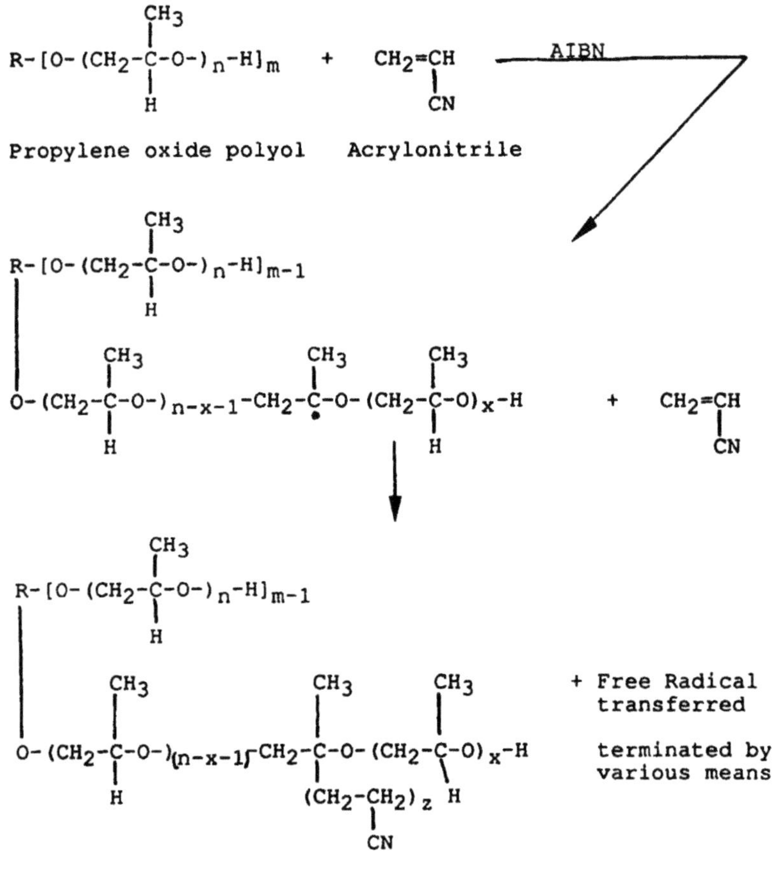

$$R-[O-(CH_2-\underset{\underset{H}{|}}{\overset{\overset{CH_3}{|}}{C}}-O-)_n-H]_m \quad + \quad CH_2=\underset{CN}{\overset{}{CH}} \quad \xrightarrow{\quad AIBN \quad}$$

Propylene oxide polyol   Acrylonitrile

$$R-[O-(CH_2-\underset{\underset{H}{|}}{\overset{\overset{CH_3}{|}}{C}}-O-)_n-H]_{m-1}$$

$$O-(CH_2-\underset{\underset{H}{|}}{\overset{\overset{CH_3}{|}}{C}}-O-)_{n-x-1}-CH_2-\underset{\underset{\bullet}{}}{\overset{\overset{CH_3}{|}}{C}}-O-(CH_2-\underset{\underset{H}{|}}{\overset{\overset{CH_3}{|}}{C}}-O)_x-H \quad + \quad CH_2=\underset{CN}{\overset{}{CH}}$$

$$R-[O-(CH_2-\underset{\underset{H}{|}}{\overset{\overset{CH_3}{|}}{C}}-O-)_n-H]_{m-1}$$

$$O-(CH_2-\underset{\underset{H}{|}}{\overset{\overset{CH_3}{|}}{C}}-O-)_{(n-x-1)}-CH_2-\underset{\underset{(CH_2-CH_2)_z}{|}}{\overset{\overset{CH_3}{|}}{C}}-O-(CH_2-\underset{\underset{H}{|}}{\overset{\overset{CH_3}{|}}{C}}-O)_x-H$$

with $(CH_2-CH_2)_z H$ and CN

+ Free Radical
  transferred

terminated by
various means

Acrylonitrile Polymer Polyol

In the case of polymer polyols prepared from ethylene oxide-
capped polyols, grafting may preferentially occur on the ethylene oxide units, as was demonstrated by Kahrs and Zimmerman

(68), who grafted vinyl esters onto poly(alkylene oxide)s. A
number of ethylenically unsaturated monomers such as styrene,
vinyl chloride, vinyl acetate, acrylonitrile, n-vinyl pyrrolidone,
vinyl propionate, ethyl acrylate, methyl methacrylate, and so
on, as well as copolymers, have been investigated (66). All
dispersions were milky or opaque in appearance, with separation
into two-phase resulting in a few instances.

Early studies indicated that acrylonitrile was the preferred
monomer both for dispersion stability through high degrees of
grafting and for property improvement. These polymer polyols
were opaque, orange/yellow-colored liquids with viscosities that
depended on the starting polyol and ranged between 1500 and
4500 cP. The use of polyacrylonitrile polymer polyols was somewhat limited in that, in flexible foam slab stock, there was a
strong tendency for marked discoloration in the center of the
"bun" where the foam was subjected to high temperatures, 150°C
to 170°C, for extended periods of time. The discoloration,
which did not have a deleterious effect on physical properties,
was thought to be related to the internal cyclization of polyacrylonitrile in which a chromophoric, conjugated, ladder-type polymeric species is formed (69).

To alleviate the discoloration problem, styrene/acrylonitrile
copolymer polymer polyols were developed (70). At styrene/
acrylonitrile ratios of 40/60 or greater with respect to styrene
content, opaque, white polymer polyols were obtained with satisfactory light, cream-colored products obtained at a 30/70 ratio. These polymer polyols had viscosities that increased with
increasing styrene content. At a 20 percent copolymer content
in a 3000 molecular weight poly(propylene oxide) triol, 30/70
and 40/60 ratio copolymers had viscosities of 1400 cP at 23°C,
whereas 50/50, 59/41, and 67/33 ratio copolymers had viscosities
of 2100, 19,500, and 35,200 cP, respectively. The viscosity increase was related to particle-to-particle association caused by
less grafting as the acrylonitrile content decreased. This resulted in relatively large, less protected particles that could
associate with each other, producing a concomitant viscosity increase. Dynamic mechanical properties were used to demonstrate modulus improvement in the rubbery region, as well as
effectively no change in the elastomer's glass transition temper-

ature and the high-temperature relaxation, T_g, of the polymer
particles that are incompatible with the elastomeric matrix.

Polymer polyols based on both polyacrylonitrile and styrene
acrylonitrile copolymers grafted both to poly(propylene oxide)
and to ethylene oxide-capped poly(propylene oxide) polyols are
commercial products (71). They are used in elastomers and
foams for a variety of applications that take advantage of the
improved load-bearing characteristics imparted by the polymer
fillers. Uses include microcellular polyurethane elastomers (72),
integral-skin polyurethane foams (73), and high-resiliency automotive seating foams (74). Polymer polyols are also used in
photocopolymerizable cycloaliphatic epoxide coating systems as
tougheners and property enhancers (75). After exposure to
ultraviolet light and cationic cocuring with the epoxides, the
polymer polyols lose opacity. In addition to excellent clarity
and mechanical properties, the coatings have good water resistance.

V. RADIATION CHEMISTRY

Exposure of poly(alkylene oxide)s to high-energy radiation,
such as electrons from electron beam or gamma radiation from
cobalt-60 sources, will alter the structure of the polymers. The
main structure alterations involved are degradation, chain scission, and crosslinking. The relative importance of these alteration processes depends on a variety of factors, such as physical state (solid or solution state), dosage, oxygen concentration,
and water or moisture content.

One of the earliest reports that poly(ethylene oxide) would
crosslink was by Charlesby in 1954 (76, 77). Studies in the
1950s were more concerned with the crosslinking reactions, later, commercial interest developed in the scission reactions,
which altered the polymers in such a manner that they had
marked improvements in such solution properties as shelf life
and shear stability that were caused by changes in molecular
weight and molecular weight distribution (78). Recently, interest appears to have centered around crosslinked poly(ethylene
oxide) as a biocompatible, semipermeable membrane that might
be used for controlled-release drug-delivery systems and blood
filtration as in dialysis and newer techniques. Dennsion (79)
points out that there are several advantages to such membranes,
one of the most important being the apparently low protein-

binding characteristics of the polymer. This allows investigators to assume that there are no enthalpic interactions between solute molecules and membrane when investigating diffusive transport. Radiation crosslinking of the polymer also offers the opportunity to investigate crosslinked networks as gels that are markedly different in structural properties, but virtually identical in chemical properties. Certain effects obtained from the transport studies suggest that it should be possible to structurally design and synthesize networks that have a very sharp size selectivity.

When crystalline or amorphous poly(ethylene oxide) is ir- radiated in bulk, scission is the main result. In marked con- trast, when aqueous solutions of the polymer are irradiated, crosslinking is the predominant process. King's investigation (78) of the solid-state irradiation with a cobalt-60 source and a Van de Graaff generator pointed out that marked changes in viscosity, molecular weight, and molecular weight distribution were related to dosage, oxygen concentration, and particle size. For example, when a 200,000 viscosity-average molecular weight polymer was subjected to dosages of 0.05 and 0.5 Mrad in air, the viscosity decreased from 680 cP for unradiated polymer to 70 and 12 cP, respectively (viscosities were for 5 percent aque- ous solutions). When dosed to the same levels in vacuum, the decrease was only to 550 and 125 cP, respectively. It was noted that the pituitous nature of aqueous polyoxyethylene solutions decreased markedly after radiation. This suggests that high- molecular-weight tails in the distributions were removed or markedly altered.

In aqueous solution, crosslinking is the predominant reac- tion process. This is related to the products of water radioly- sis interacting with the polyoxyethylene chain, and this was examined by a number of investigators (79–86). Studies have also been carried out in t-butyl alcohol/water mixtures (87), chloroform (86), and dioxane (88).

When water undergoes radiolysis, a number of reactions, which are summarized below, can take place.

$$H_2O \xrightarrow{\text{hv}} H\cdot \; + \; HO\cdot \; + \; (e_{aq})^-$$

$$HO\cdot \; + \; HO\cdot \; \longrightarrow \; H_2O_2$$

$$H\cdot \; + \; HO\cdot \; \longrightarrow \; H_2O$$

$$H\cdot \; + \; H\cdot \; \longrightarrow \; H_2$$

$$(e_{aq})^- + H_2O_2 \longrightarrow \cdot OH + OH^-$$

$$OH\cdot + OH^- \longrightarrow H_2O + O\cdot^-$$

Some of these compounds can interact with poly(ethylene oxide) through a hydrogen-abstraction mechanism, and others can react or interact, with both degradation and crosslinking resulting. Both the activated hydroxyl group, HO·, and the activated oxygen atom, O·, can cause hydrogen abstraction from the polymer backbone.

$$-CH_2-CH_2-O- \ + \ O\cdot \ \rightarrow \ -CH_2-C\cdot H-O- \ + \ OH^-$$

$$-CH_2-CH_2-O- \ + \ \cdot OH \ \rightarrow \ -CH_2-C\cdot H-O- \ + \ H_2O$$

The poly(ethylene oxide) molecules with free-radical species on their background can combine to form a crosslinked system.

$$2 \ -CH_2-C\cdot H-O- \ \longrightarrow \ -CH_2-CH-O-$$
$$\text{etc.}$$
$$-CH_2-CH-O-$$

Depending on various factors, such as molecular weight and solvent quality, crosslinking can be intermolecular or intramolecular (89–92), with the latter favored in salt solutions that cause the molecules to be in a tightly coiled configuration (93).

Interactions with the hydrated electrons or hydrogen peroxide lead to molecular weight degradation reactions.

$$-CH_2-CH_2-O-CH_2-C\cdot H-O- \ + \ H_2O_2 \ \rightarrow \ \cdot OH \ +$$
$$-CH_2-CH_2-OH \ + \ OCH-CH_2-$$

$$-CH_2-CH_2-O-CH_2-C\cdot H-O- \ + \ (e_{aq})^- \ \rightarrow \ H_2O \ +$$
$$-CH_2-CH_2-O^- \ + \ CH_2=CHO-$$

Although a range of degradation reactions has been proposed to take place in the presence of oxygen (82), it appears that at high dosage rates, such as those achieved with electron-beam generators, the oxygen is rapidly depleted and cross-linking reactions predominate. Dennison (79) found no measurable difference in molecular weight distribution between samples irradiated in an inert atmosphere (nitrogen) and in the presence of oxygen.

Minkova and coworkers (94) irradiated high-molecular-weight, 2.5×10^6, poly(ethylene oxide) with gamma rays from a cobalt-60 source using dosages of 1 to 15 mrad. Optical and electron microscopy indicated that homogeneous networks that originated by means of both intramolecular and intermolecular crosslinking were formed. Larger chain segments that formed by chain scission were concurrently grafted to the network during the radiation process. The degree of crystallinity decreased as irradiation dosage (i.e., degree of crosslinking) increased. Both enthalpy of melting and melting point had marked decreases with increasing dosage up to 5 Mrad. From 5 to 8 Mrad, these parameters continued decreasing, but at a decreased rate—about three times slower. At dosages of 8 to 15 Mrad, the parameters did not change. The changes are related to the expected reduction in crystallinity with increasing crosslink density and to free-energy changes arising from an entropic origin in both crystalline and amorphous areas.

A technique for immobilizing polyethylene glycol (PEG) on poly(ethylene terephthalate) film has been described by Gombotz and coworkers (95). Such bonding or immobilization places the hydrophilic, nonionic polymeric glycol onto a solid, tough, flexible substrate, and in this form it has special utility in the chemical, cosmetic, and biomedical industries. Cleaned poly-(ethylene terephthalate) is first treated by exposure to a radio frequency glow discharge plasma of either allyl amine or allyl alcohol to form amino or hydroxyl groups on the surface. These functional groups are then activated with cyanuric chloride. In the case of film exposed to an allyl amine plasma, the treated film is reacted first with the cyanuric chloride and then with an amine-terminated polyethylene glycol (see Section III).

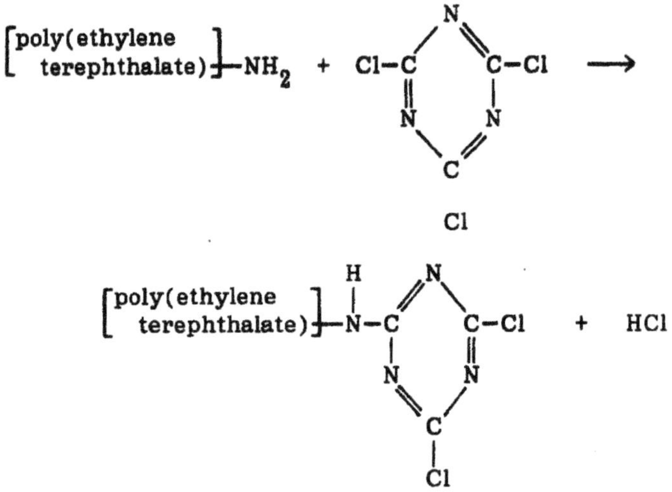

$$
\left[\begin{array}{l}\text{poly(ethylene}\\ \quad\text{terephthalate)}\end{array}\right]\!-NH_2 \;+\; Cl-C\!\!\!\overset{\displaystyle N}{\underset{\displaystyle N}{\diagup\diagdown}}\!\!\!C-Cl \;\longrightarrow
$$

$$
\left[\begin{array}{l}\text{poly(ethylene}\\ \quad\text{terephthalate)}\end{array}\right]\!-\overset{\displaystyle H}{N}\!-C\!\!\!\overset{\displaystyle N}{\underset{\displaystyle N}{\diagup\diagdown}}\!\!\!C-Cl \;\;+\;\; HCl
$$

(I) cyanuric chloride-activated
poly(ethylene terephthalate) film

$$
(I) \;+\; H_2N-(PEG)-NH_2 \;\rightarrow\; (I)-\overset{\displaystyle H}{N}-(PEG)-NH_2 \;+\; HCL
$$

substrate with
immobilized PEG

In the case of poly(ethylene terephthalate) film exposed to allyl
alcohol, the film is treated with a butyllithium solution to form
the nucleophilic alkoxide ion. This film is then derivatized
with the cyanuric chloride. The hydroxyl end groups of polyethylene glycol are reacted with butyllithium to form the alkoxide derivative and then with the cyanuric chloride-activated
film.

$$
\left[\begin{array}{l}\text{polyethylene}\\ \quad\text{terephthalate}\end{array}\right]\!-OH \;+\; BuLi \;\rightarrow\; \left[\begin{array}{l}\text{polyethylene}\\ \quad\text{terephthalate}\end{array}\right] O^-Li^+
$$

alkoxide A

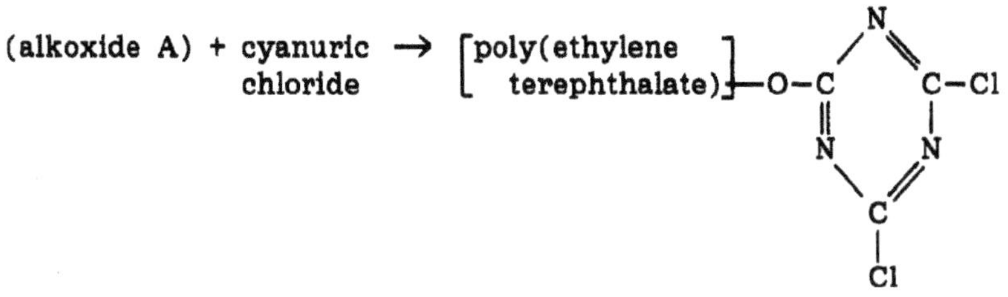

(alkoxide A) + cyanuric → [poly(ethylene terephthalate)]—O—C ... C—Cl

(II) cyanuric chloride-activated
poly(ethylene terephthalate) film

$$(II) + Li^{+-}O(PEG)O^-Li^+ \rightarrow (II)-O(PEG)O^-Li^+ + LiCl$$

substrate with
immobilized PEG

Other techniques for preparing substrates with covalently attached ethylene oxide units are referenced.

VI. ELASTOMERS, CURING TYPES

This section is concerned with poly(alkylene oxide)s having sufficiently high molecular weight to be gumstocks rather than the liquid polyols used to prepare polyurethanes, polyureas, and polyurethane-ureas that are discussed elsewhere. Gumstocks such as those described herein can be compounded, shaped, and cured by means of chemical reaction. As an addendum to the information in this section, it should be noted that BF Goodrich has signed a letter of intent to sell its elastomer business to Zeon Chemicals USA, Inc., a subsidiary of Nippon Zeon, Tokyo (96). Poly(propylene oxide) and poly(epichlorohydrin) elastomers are available from Zeon, whose corporate headquarters and R&D center are located in Louisville, Kentucky, rather than at BF Goodrich.

A. Poly(propylene oxide) Elastomers

Although early investigators postulated that polyethers such as poly(propylene oxide) should yield excellent elastomers because of the chain flexibility that would be contributed by the ether oxygen, propylene oxide could not be polymerized to a sufficiently high molecular weight for rubbery characteristics. This situation changed when coordination catalysts that involved the use of organoaluminum compounds in combination with water and acetyacetone were discovered (see Chapter 4, Section III). These catalysts, which were discovered and reported in 1960 by Vandenberg (97), caused propylene oxide to polymerize to sufficiently high molecular weights that the polymer had rubbery characteristics. From studies based on this early work, a poly(propylene oxide) elastomer, PAREL^{TM}, was developed (98) and commercialized in 1972. The current commercial elastomer (99), PAREL^{TM} 58, is made vulcanizable by copolymerizing propylene oxide and allyl glycidyl ether that originally was present in a 94/6 weight ratio. The resulting gumstock has the following average structure:

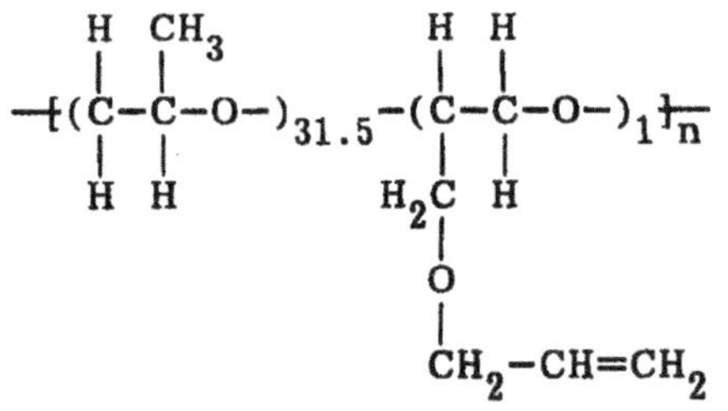

Commercial poly(propylene oxide) gumstock

This structure has an average molecular weight between crosslink sites, M_c, of about 1800. The gumstock has a slight odor, is white to light amber in color, has a molecular weight of about 1.5×10^6, and is supplied in slab form.

Essentially simultaneously as the above work was being carried out, Bailey (100) developed similar poly(propylene oxide) gumstocks, but used butadiene monoepoxide and similar compounds including allyl glycidyl ether as the comonomer to place pendant unsaturation onto the macromolecular backbone. This copolymer had the following structure, with the parameter a unspecified and n sufficiently high to produce gumstocks rather than low-viscosity liquids.

$$-[(\overset{\overset{\displaystyle H}{|}}{\underset{\underset{\displaystyle H}{|}}{C}}-\overset{\overset{\displaystyle CH_3}{|}}{\underset{\underset{\displaystyle H}{|}}{C}}-O-)_a-(\overset{\overset{\displaystyle H}{|}}{\underset{\underset{\displaystyle HC}{|}}{C}}-\overset{\overset{\displaystyle H}{|}}{\underset{\underset{\displaystyle H}{|}}{C}}-O-)]_n-$$
$$\underset{\underset{\displaystyle CH_2}{\|}}{}$$

Copolymers with ethylene oxide and other epoxides were also developed (101).

Vulcanization is accomplished by reaction of sulfur-based cure packages with the pendant unsaturation, and, if desired, the elastomer can be reinforced with carbon black. Although it would appear that peroxides could also effect crosslinking, in addition to crosslinking through the unsaturation, they also cause chain scission, which is the predominant reaction. Since propylene oxide contains an asymmetric carbon atom, various stereoisomers can form, as well as head-to-head, tail-to-tail, and the expected head-to-tail units that can exist in the polymer chain. As commercially produced, this gumstock has very little crystallinity by design, though highly crystalline polymers have been made (102). Mark-Houwink equations that describe the relationship between intrinsic viscosity and molecular weight have been determined (103).

$$[\eta] = 1.12 \times 10^{-4} M_w^{0.77} \quad (25°C, \text{ benzene})$$

$$[\eta] = 1.97 \times 10^{-4} M_w^{0.67} \quad (46°C, \text{ hexane})$$

Selected properties of the poly(propylene oxide) elastomer are given in Table 1, along with information about the effect of elevated-temperature aging. From these data, it is apparent that the cured elastomer withstands high temperatures very well. The excellent balance of upper and lower use temperatures and fair hydrocarbon resistance make this elastomer an alternative to natural rubber in automotive parts that are used in high-temperature, dynamic applications such as motor mounts, belts, and insulators. Added advantages over natural rubber include good ozone resistance and resistance to aging at high temperatures. Because of the good low-temperature properties, the use of plasticizers is seldom required. When it is necessary to use plasticizers to decrease compound viscosity or decrease

TABLE 1   Properties of Poly(propylene oxide) Elastomers Before
and After Thermal Aging (From Refs. 99, 104.)

Property	Before thermal aging	After thermal aging 70 hr at 150°C
Glass transition temperature, °C	−75	—
Use Temperature		
Low, °C	−60	—
High, °C	136	—
100% Modulus, psi (MPa)	550 (3.8)	660 (4.5)
200% Modulus, psi (MPa)	1190 (8.2)	1450 (10.0)
Elongation, %	370	210
Tensile strength, psi (MPa)	1930 (13.3)	1580 (10.9)
Hardness, Shore A	70	73
Bashore resilience, %	52	—
Low-temperature stiffness, $T_{10,000}$, ASTM D1053, °C	−58	—

hardness of highly filled systems, up to 30 phr of naphthenic
or aromatic oils such as Sundex^{TM} 790 (Sun Petroleum Products
Co.) may be used.

Berta and Vandenberg (104) have reviewed the background
of the commercial elastomers and describe methods of production, compounding including cure packages, and properties of
the cured elastomers.   Commercially available literature (99)
details processing, toxicology, filler compounding, and cured
properties.   Other investigators (105) have examined the elasticity of model poly(propylene oxide) networks and obtained results that were in general agreement with the theory of elasticity according to Flory.   However, their data with highly crosslinked networks, $M_c \sim 725$, could not be satisfactorily described
by more recent theories.   This was felt to be due to the marked

chemical modification that takes place at high crosslink densities and to the non-Gaussian character of the short network chains studied.

A partially stereoregular poly(propylene oxide) homopolymer was fractionated into a 0°C isooctane insoluble, K-polymer, and a soluble, D-polymer, fractions (106). The K-polymer was high-molecular-weight, stereoregular poly(propylene oxide) and represented from 7 to 27 percent of the whole polymer, with the amount of K-polymer depending on polymerization time and temperature and on catalyst concentration. The D-polymer contained no unsaturation and was primarily a cyclic, low-molecular-weight (~1000) oligomer. Linear polymer contained in this fraction had hydroxyl end groups.

B. Polyepichlorohydrin Elastomers (107—109)

Epichlorohydrin was first successfully polymerized to high molecular weight with the coordination catalysts mentioned above (97). The homopolymer of epichlorohydrin,

$$—(CH_2-CH-O-)_n—$$
$$|$$
$$CH_2Cl$$

and a 1:1 mole ratio (68/32 weight ratio) of epichlorohydrin and ethylene oxide copolymer,

$$—(CH_2-CH-O-CH_2-CH_2-O-)_n—$$
$$|$$
$$CH_2Cl$$

are important commercial products marketed as Hydrin^{TM} 100 and 200, respectively (107). A third commercial polymer, Hydrin^{TM} 400, is a terpolymer of epichlorohydrin, ethylene oxide, and an unsaturated monomer. These amorphous gumstocks are easy to process and, when formulated, are easy to extrude and calendar. Crosslinking is accomplished through the chlorine atom by means of imidazolines, diamines, ureas, and ammonium salts (110—112). Such compounds react with the chloromethyl side groups, forming hydrochloric acid as a by-product. High degrees of crosslinking are obtained by use of a suitable acid acceptor such as metal oxides, zinc stearate, calcium stearate, etc., to remove the acid as it is formed. The elastomers can be reinforced with carbon black and silicas, and sulfur is used to improve tensile properties of such reinforced products. The unsaturation in Hydrin^{TM} 400 offers another dimension to cure in that it can be blended with and cocured with conventional gumstocks such as nitrile and styrene-butadiene gums. Details regarding compounding, formulation, and cured properties can be found in the general references given at the start of this section (107–109).

When crosslinked, these elastomers have superior heat-aging and ozone resistance, a high degree of resistance to swelling in a number of liquids including petroleum products, very low oxygen and other gas permeability, and good chemical resistance to acids and bases. Since the chlorine content is quite high, the elastomers have fair-to-good flame-resistance properties. Overall, the polyepichlorohydrin elastomers have a blend of the good characteristics of nitrile and neoprene elastomers. Table 2 contains some of the physical attributes of the elastomers.

Blends of polyepichlorohydrin and poly-epsilon-caprolactone have been investigated. The blends are miscible; but, at levels above about 30 percent polylactone, phase separation that is caused by the strong crystallization tendency of the polylactone takes place (113). The relation

$$(1/Tg_{1,2}) = (W_1/202 \text{ K}) + (W_2/257 \text{ K})$$

where $Tg_{1,2}$ is the glass transition temperature of the blend and W_1 and W_2 are the weight fractions of poly-epsilon-caprolactone and polyepichlorohydrin, respectively, adequately describes the glass transition temperature of the blends. Polyepichlorohydrin has been found to be miscible with a number of methacrylates and acrylates (114, 115). In the case of poly(n-butyl acrylate), only partial miscibility was obtained. The commercial epichlorohydrin/ethylene oxide copolymer was miscible only with poly(methyl acrylate) and not with the higher acrylates. The same was true for the miscibility of polyacrylates with poly(ethylene oxide). It was concluded that the exothermic interactions between epichlorohydrin and ethylene oxide units in the copolymer are sufficiently strong to preclude miscibility with all acrylates except poly(methyl acrylate). The epichlorohydrin elastomer has been added to poly(vinyl chloride) and to polysulfones to improve impact resistance (116).

Epichlorohydrin elastomers are used as belts, gaskets, oil seals, paper-mill rolls, printing rolls and blankets, oil-field

TABLE 2   Properties of Polyepichlorohydrin Elastomers (From Refs. 107, 113, 114.)

Polyepichlorohydrin	
Glass transition temperature	$-15°C$
Gehman torsion stiffness, T_{100}, ASTM D1053	$-26°C$
Use temperature	
Low	$-40°C$
High	$163°C$
Tensile strength	2500 psi (17.2 MPa)
Percent elongation	450%
Hardness, shore A	50—85
Epichlorohydrin/Ethylene Oxide Copolymer	
Gehman torsion stiffness, T_{100}, ASTM D1053	$-41°C$
100% modulus	450 psi (3.1 MPa)
Tensile strength	1920 psi (13.2 MPa)
Elongation	630%
Hardness, shore A	85

specialties, hoses for handling petroleum, molded mechanical goods, adhesives, coated fabrics, diaphragms, wire and cable jackets, and collapsible fuel containers.   Applications in the automotive area take advantage of the excellent heat-, oil-, and ozone-resistance characteristics.

VII.   OXIDATION AND THERMAL DEGRADATION

Polyethers are susceptible to oxidation and usually are stabilized with an antioxidant.   The oxidative attack of polyoxyethylene proceeds by an autooxidation mechanism that involves intrachain hydroperoxide formation, which decomposes and causes chain

cleavage (117–119). The process is accelerated by ultraviolet
light, strong acids, heavy metal ions (120), and certain oxidiz-
ing agents. McGary (117) investigated the attack in aqueous
solution and found allyl, ethyl, and isopropyl alcohol to be effective stabilizers, with 2–5 percent isopropanol being particularly effective. In ethanol solution, no degradation was observed, and polymers precipitated from ethanol did not undergo
chain scission at temperatures above 150°C (121). The investigation, which involved degradation in solution and in bulk,
found that free-radical inhibitors had little effect, but the antioxidant, 2,4-di-t-butyl cresol, did retard decomposition in bulk
and ethanol was effective in benzene solution. Additive systems
consisting of 2,2'methylenebis(4-methyl-6-t-butyl phenol) and
2,6-di-t-butyl p-cresol (122) and phenothiazine or alkylpheno-
thiazine and certain boron compounds (123) were found to be
particularly effective stabilizers. Carbamates (124) and various
phenols and cresols (125) have also been used. Other oxidative and thermal stabilizers for polyoxyethylene include pipera-
zine and derivatives (126) and 2-mercaptoimidazoline and related
compounds (127). Goldschmidt and coworkers (128) indicate
that mixtures of aliphatic amines and phenolic compounds with
at least two hydroxyl groups on the aromatic ring are good oxidative stabilizers. The aromatic compounds are the primary
inhibitors and include hydroquinone, 1,2- and 1,4-dihydroxy
naphthalene, di-t-butyl resorcinol, and similar compounds. The
aliphatic amines are secondary inhibitors and include ethanol-
amine, isopropanolamine, ethylenediamine, etc.

Poly(ethylene oxide) can also be degraded by mechanical
action such as high shearing forces, with both shear level and
duration of shearing force being important variables (129–132).
As molecular weight increases, the effect is markedly more no-
ticeable. Even the low shear forces encountered in capillary
viscometry can have an effect and cause errors to arise in determination of specific viscosity (133). Shear degradation can
cause pronounced broadening of the molecular weight distribution and can preclude use of gel permeation chromatography as
an investigative tool. In certain solvents, such as tetrahydrofuran, large differences in intrinsic viscosity (12.4 to 1.7 dl/g)
have been noted with relatively, but markedly, smaller changes
in other solvents such as 50/50 by volume water/ethanol mixtures (134). Other investigators have had good results when
2-ethoxyethanol at 80°C was used as the elution solvent (135).
Investigations of photoirradiation in deaerated systems led to
degradation with the formation of methane, ethane, and carbon

dioxide under acidic conditions and of hydrogen under basic or neutral condtions (136). Methods of improving stability by polymerization-catalyst selection have been studied (137). The stages of autooxidation have also been under investigation (138). Thermal decomposition of polyoxyethylene along and in combination with methyl methacrylate is another interesting way to examine the degradation of polyoxyethylene and other polyethers (139, 140).

Polyoxyethylenes have been reacted with different peroxycarbonates, which leads to functionalization—i.e., free-radical acetonylation or formylmethylation (141). Chains of 400 to 600,000 molecular weight can be functionalized by using only small amounts of the peroxycarbonate. Difficulty in functionalizing increases with increasing molecular weight, because of the difficulty in fluidizing the high-molecular-weight compounds.

One of the deficiencies of polyoxypropylene is oxidative stability. Notice of the tertiary hydrogen atom that is alpha to an ether linkage in the molecules and classical ether oxidation chemistry makes the instability readily apparent. The data of St. Pierre and Price (142)—who bubbled oxygen into polyoxypropylene that was held in a boiling water bath (i.e., ~100°C) and determined acidity, peroxide content, and specific viscosity—are plotted in Figure 1. It is readily apparent that after a short induction period, both peroxide and acidic content increase in an essentially linear manner. Specific viscosity or molecular weight decreases after a small initial increase that may be related to association of carboxy groups. These investigators also studied thermal degradation at 270°C in a nitrogen atmosphere under neutral, basic, and acidic conditions and in air under neutral conditions. Acidic, p-toluene sulfonic acid, conditions were the most severe, and basic, potassium hydroxide, conditions had the least effect. Neutral conditions in nitrogen were noticeably more gentle than in air, which implies that oxidation was the predominant route to molecular weight degradation.

Only small amounts, 200 to 300 ppm, of an antioxidant such as 2,6-di-t-butyl-4-methyl phenol (DBP) can be sufficient to stabilize polyoxypropylene during storage (143). However, when products are manufactured from polyoxypropylene polyols, high temperatures for prolonged times can be encountered. For elastomers and molded flexible foam products where temperature build and maintenance is not a problem, 500 to 1200 ppm DBP will usually suffice as a stabilizer. However, in slab-stock foam that is produced in bun form, temperatures in the central

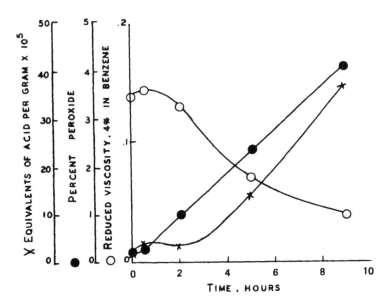

FIG. 1 Degradation of polyoxypropylene in an oxygen environment at ~100°C.   (From Ref. 142.)

portion of the bun can be in excess of 150°C and can remain at this temperature for several hours.   In such systems, larger amounts of stabilizer—1000 to 5000 ppm DBP—are required.   In the case of rigid foams, which use polyols that have only low degrees of propoxylation and, thus, have little polyether character, little or no antioxidant is needed.

It has been shown that the vacuum thermal degradation of poly(ethylene oxide) is a first-order reaction at 50°C to 90°C with hydroperoxide concentrations of 0.7 to 1.35 mol/kg (144). The main degradation products were explained as arising from transformation of an intermediate epoxide ring through either a radical or a heterolytic mechanism.

REFERENCES

1. Annual Book of ASTM Standards, Part 36, American Society for Testing and Materials, Philadelphia, 1980.

2. S. L. Wellons, M. A. Cary, and D. K. Elder, Anal. Chem. 52:1374 (1980).

3. W. B. Satkowski, S. K. Huang, and R. L. Liss, in Nonionic Surfactants (M. J. Schick, ed.), Marcel Dekker, New York, 1967, p. 142.

4. R. A. Reck, in Fatty Acids in Industry (R. W. Johnson and E. Fritz, eds.), Marcel Dekker, New York, 1988, p. 688.

5. M. J. Astle, B. Schaeffer, and C. O. Obenland, J. Am. Chem. Soc. 77:3643 (1955).

6. J. D. Malkemus and J. I. Swan, J. Am. Oil Chemists Soc. 34:342 (1957).

7. R. H. Kienle and G. P. Whitcomb, U.S. Patent No. 2,473,798 (1949).

8. A. N. Wrigley, F. D. Smith, and A. J. Stirton, J. Am. Oil Chemists Soc. 34:39 (1957).

9. S. Sussman, Ind. Eng. Chem. 38:1228 (1946).

10. H. Bennett, U.S. Patent No. 2,275,494 (1942).

11. W. Schlutter, French Patent 1,031,835 (1955).

12. K. Nagase and K. Sakaguchi, Kogyo Kagaku Zasshi 64: 1035 (1961).

13. A. N. Wrigley, F. D. Smith, and A. J. Stirton, J. Oil Chemists Soc. 36:34 (1959).

14. K. M. Chen and H. J. Liu, J. Appl. Polymer Sci. 34: 1878 (1987).

15. F. E. Critchfield, J. V. Koleske, and R. A. Dunleavy, Rubber World 163:51 (August 1971).

16. C. G. Seefried, Jr., J. V. Koleske, and F. E. Critchfield, J. Appl. Polymer Sci. 19:2503 (1975).

17. J. R. Craven, C. V. Nicholas, R. Webster, D. J. Wilson, R. H. Mobbs, G. A. Morris, F. Heatley, C. Booth, and J. R. M. Giles, Br. Polymer J. 19:509 (1987).

18. J. R. Craven, R. H. Mobbs, C. Booth, and J. R. M. Giles, Makromol. Chem., Rapid Comm. 7:81 (1986).

19. R. H. Zdrahala, E. M. Firer, and J. F. Fellers, J. Polymer Sci., Polymer Chem. Ed. 15:689 (1977).

20. B. Bogdanopv, M. Michailov, and A. Popov, Acta Polym. 39 (7):385 (1988).

21. T. Wodka and M. Danielewicz, Acta Polym. 39 (10):539 (1988).

22. W. A. Feld, F. W. Harris, and B. Ramalingam, ACS Polymer Preprints 21 (1):215 (March 1981).

23. J. A. Brydson, Plastics Materials, D. Van Nostrand, Princeton, New Jersey, 1966, p. 576.

24. H. Lee and K. Neville, Handbook of Epoxide Resins, McGraw Hill, New York, 1967.

25. M. K. Purgett, W. Deits, and O. Vogel, J. Polymer Sci., Polymer Chem. Ed. 20:2477 (1982).

26. T. Nishikubo, T. Ichijyo, and T. Takoaka, Nippon Kagaku Kaishi 35:871 (1973); C.A. 78: 160375 (1973).

27. T. D. N'Guyen, A. Deffieux, and S. Boileau, Polymer 19: 423 (1978).

28. T. Nishikubo, T. Iizawa, Y. Sugawara, and T. Shimokawa, J. Polymer Sci., Polymer Chem. Ed. 24:1097 (1986).

29. T. Nishikubo, T. Shimokawa, T. Fujii, T. Iizawa, Y. Harita, and M. Koshiba, J. Polymer Sci., Part A: Polymer Chem. 26:2881 (1988).

30. I. Vulic, T. Okano, S. W. Kim, J. Polymer Sci., Part A: Polymer Chem. 26:381 (1988).

31. A. Suekane, T. Masuda, and H. Kawakami, Japan Kokai Tokkyo Koho JP 63/210,128 (1988); C.A. 110: 76274 (1989).

32. K. S. Kazanskii and N. V. Ptitsyna, Makromol. Chem. 190:255 (1989).

33. A. D. Smith, Tenside Deterg. 20 (1):23 (1983).

34. R. Duran and C. Strazielle, Macromolecules 20:2853 (1987).

35. R. T. K. Cornwell, in Industrial Gums (R. L. Whistler and J. N. BeMiller, ed.), Academic Press, New York, 1959, p. 597.

36. H. M. Spurlin, in Cellulose and Cellulose Derivatives, Part II (E. Ott, H. M. Spurlin, and M. W. Grafflin, eds.), Interscience Publishers, New York, 1954, p. 673.

37. H. Dreyfus, Br. Patent 166,767 (1921); CA: 16:830 (1922).

38. E. D. Klug and H. G. Tennent, U.S. Patent No. 2,172,109 (1939).

39. D. R. Erickson, U.S. Patent No. 2,469,764 (1949).

40. W. Brown, Arkiv For Kemi 18 (16):227 (1961).

41. W. Brown, D. Henley, and J. Ohman, Makromol. Chem. 64:49 (1963).

42. F. Friedberg, W. Brown, D. Henley, and J. Ohman, Makromol. Chem. 66:168 (1963).

43. W. Brown and D. Henley, Makromol. Chem. 75:179 (1964).

44. H. Elmgren and S. Norrby, Makromol. Chem. 123:265 (1969).

45. D. Donescu, K. Gosa, I. Disconescu, N. Carp, and M. Mazare, Colloid and Polymer J. 258:1363 (1980).
46. W. Brown and K. Chitumbo, Chemica Scripta 2:88 (1972).
47. I. Jullander, Svensk Papperstidn. 55:197 (1952).
48. S. Sonnerskog, Svensk Papperstidn. 48:413 (1945).
49. J. L. Schick, U.S. Patent No. 2,538,051 (1951).
50. S. Okimasu, J. Agricul. Chem. Soc. Japan 30:36 (1956).
51. T. Okujama, Protein Nucleic Acid Enzyme 15:47 (1970).
52. S. N. Danilov and E. A. Plisko, Zhur. Obshch. Khim. 28:2217 (1958).
53. M. Weber and R. Stadler, Polymer Preprints, American Chem. Soc., Div. Poly. Chem. 30 (1):109 (1989).
54. J. M. Lee and J. C. Winfrey, U.S. Patent No. 3,236,895 (1966).
55. E. L. Yeakey, U.S. Patent No. 3,654,370 (1972).
56. E. L. Martin, U.S. Patent No. 2,359,867 (1944).
57. K. B. Sellstrom, Adhesives Age 31:30 (June 1988).
58. T. E. Bishop, C. J. Coady, J. M. Zimmerman, G. K. Noren, and C. E. Fisher, European Patent Publication 209,641 (1987).
59. E. C. Y. Nieh, U.S. Patent No. 4,710,362 (1987).
60. G. J. McCollum, R. L. Scriven, R. M. Christenson, G. W. Mauer, and R. R. Zwack, U.S. Patent No. 4,711,917 (1987).
61. J. C. Craponne and J. G. Ecully, U.S. Patent No. 4,717,763 (1988).
62. R. J. Steltenkamp and M. A. Camara, U.S. Patent No. 4,714,559 (1987).
63. R. J. McCready and J. A. Tyrell, U.S. Patent No. 4,711,948 (1987).
64. B. Masar, V. Janout, and P. Cefelin, Makromol. Chemie 189:2323 (1988).
65. Y. Imai, M. Kajiyama, S. Ogata, and M. Kakimoto, Polymer J. (Japan) 16 (3):267 (1984).
66. W. C. Kuryla, F. E. Critchfield, L. W. Platt, and P. Stamberger, J. Cellular Plastics 2:84 (March 1966).
67. P. Stamberger, U.S. Patent No. 3,304,273 (1967).
68. K-H. Kahrs and J. W. Zimmerman, Makromol. Chem. 58:75 (1962).
69. R. C. Houtz, Textile Res. J. 20:786 (1950).
70. F. E. Critchfield, J. V. Koleske, and D. C. Priest, Rubber Chem. Tech. 45:1467 (1972).

71. Union Carbide Corp., Specialty Chem. Div., Brochure SC-650, Performance Polyether Polymer Polyols (Feb. 1988).

72. R. A. Dunleavy, J. Elastoplastics 2:1 (Jan. 1970).

73. R. D. Whitman, Plastics Tech. 16:47 (Feb. 1970).

74. W. Patten and D. C. Priest, J. Cellular Plastics 8:134 (May/June 1972).

75. J. V. Koleske, U.S. Patent No. 4,593,051 (1986).

76. A. Charlesby, Nature 173:679 (1954).

77. A. Charlesby, Atomic Radiation and Polymers, Pergamon Press, New York (1960).

78. P. A. King, in Irradiation of Polymers, Advances in Chemistry Series No. 66, Am. Chem. Soc., Washington, D.C., 1967, p. 275.

79. K. A. Dennison, Radiation Crosslinked Poly(Ethylene Oxide) Hydrogel Membranes, Doctoral Thesis, Massachusetts Institute of Technology, Cambridge, 1986.

80. R. A. Van Brederode, F. Rodriguez, and G. G. Cocks, J. Appl. Poly. Sci. 12:2097 (1968).

81. C. Decker, M. Vackerot, and J. Marchal, Comptes Rendu. Acad. Sc. Paris 261:5104 (1965).

82. C. Crouzet and J. Marchal, Makromol. Chem. 166:99 (1973).

83. C. Decker and J. Marchal, Makromol. Chem. 166:155 (1973).

84. P. A. King and J. A. Ward, J. Polymer Sci., A-1 8:253 (1970).

85. M. S. Matheson, A. Mamou, J. Silverman, and J. Rabani, J. Phys. Chem. 77 (20):2420 (1973).

86. F. Gugumus and J. Marchal, J. Polymer Sci. Part C 16:3963 (1968).

87. J. W. Stafford, Makromol. Chem. 134:87 (1970).

88. J. W. Stafford, Makromol. Chem. 134:113 (1970).

89. U. Borgwardt, W. Schnabel, and A. Henglein, Makromol. Chem. 127:176 (1969).

90. J. W. Stafford, Macromol. Chem. 134:71 (1970).

91. W. Schnabel, Makromol. Chem. 131:287 (1970).

92. U. Grollman and W. Schnabel, Makromol. Chem. 181:1215 (1980).

93. R. A. Van Brederode and F. Rodriguez, J. Appl. Poly. Sci. 14:979 (1970).

94. L. Minkova, R. Stamenova, C. Tsvetanov, and E. Nedkov, J. Polymer Sci., Part B: Polymer Phys. 27:621 (1989).

95. W. R. Gombotz, W. Guanghui, and A. S. Hoffman, J. Appl. Polymer Sci. 37:91 (1989).

96. Anon., Chem. Eng. News 67:7 (6/5/89); E. S. Kiesche, Chem. Week 145:13 (6/14/89); Anon., Chem Week 145:5 (11/1/89).

97. E. J. Vandenberg, J. Polymer Sci. 47:486 (1960); U.S. Patent No. 3,158,580 (1964); U.S. Patent No. 3,158,581 (1964); U.S. Patent No. 3,158,591 (1964).

98. E. J. Vandenberg, U.S. Patent No. 3,728,320 (1973).

99. BF Goodrich, Brochure HM-18, PAREL^{TM} Elastomers, (Jan. 1987).

100. F. E. Bailey, Jr., U.S. Patent No. 3,031,439 (1962).

101. F. E. Bailey, Jr., U.S. Patent No. 3,417,064 (1968).

102. J. M. Baggett and M. E. Pruitt, U.S. Patent No. 2,871,219 (1959).

103. G. Allen, C. Booth, and M. N. Jones, Polymer 5:195 (1964).

104. D. A. Berta and E. J. Vandenberg, in Handbook of Elastomers: New Developments and Technology (A. K. Bhowmick and H. L. Stephens, eds.), Marcel Dekker, New York, 1988, p. 643.

105. A. L. Andrady and M. A. Llorente, J. Polymer Sci., Part B: Polymer Physics 25: 195 (1987).

106. K. Alyuruk, T. Ozden, and N. Colak, Polymer 27:2009 (1986).

107. BF Goodrich, Brochure HM-13, Hydrin^{TM}Elastomers (1987).

108. W. D. Willis, L. O. Amberg, A. E. Robinson, and E. J. Vandenberg, Rubber World 153 (1):88 (1965).

109. E. J. Vandenberg, in Kirk-Othmer Encyclopedia of Chemical Technology, Vol. 8, 3d ed., John Wiley & Sons, New York, 1979, p. 930.

110. A. E. Robinson and W. D. Willis, Belg. Patent 636,960 (1964).

111. Herclor Rubbers, Bulletin ORH-23, Hercules Inc., Wilmington, Delaware.

112. A. E. Robinson, U.S. Patent No. 3,026,270 (1962).

113. G. L. Brode and J. V. Koleske, J. Macromol. Sci.- Chem. A6 (6):1109 (1972).

114. J. G. Ceryan, in Handbook of Materials and Processes for Electronics, McGraw Hill, New York, 1972, p. 521.

115. A. C. Fernandes, J. W. Barlow, and D. R. Paul, J. Appl. Polymer Sci. 32:6073 (1986).

116. F. Wollrub, Polymer Preprints 13 (1):499 (1972).

117. C. W. McGary, Jr., J. Polymer Sci. 46:51 (1960).

118. P. Grosborne, I. S. deRoch, L. Sajus, Bull. Chim. Soc. France :2020 (1968).

119. L. Reich, S. S. Stivaln, J. Appl. Polymer Sci. 13:977 (1969).

120. H. Vink, Makromol. Chemie 67:105 (1963).

121. A. M. Afifi and J. N. Hay, Eur. Polymer J. 8:289 (1972).

122. J. E. Tyre and F. G. Willeboordse, U.S. Patent No. 3,388,169 (1968).

123. C. R. Dickey, U.S. Patent No. 3,374,275 (1968).

124. T. J. Dolce, F. M. Berardinelli, and D. E. Hudgin, U.S. Patent No. 3,144,431 (1964).

125. D. G. Leis and E. S. Stout, U.S. Patent No. 2,942,033 (1960).

126. K. Konishi, Y. Sano, and E. Ajisaka, JA 70/03,224 (1970).

127. K. Kenichi, Y. Ueda, W. Nishihama, Y. Tanino, W. Narukawi, Ger. Offen. 1,948,909 (1970).

128. A. Goldschmidt, W. T. Stewart, E. Cerrito, and O. L. Harle, U.S. Patent No. 2,687,378 (1954).

129. K. Konishi, Y. Sano, and E. Ajisaka, JA 70/03,224 (1970).

130. W. K. Asbeck and M. K. Baxter, presentation at Am. Chem. Soc. Chicago Meeting (Sep. 1958); F. E. Bailey, and J. V. Koleske, in Nonionic Surfactants: Physical Chemistry (M. J. Schick, ed.), Marcel Dekker, New York, 1987, p. 944.

131. Y. Minoura, T. Kasuya, S. Kawamura, and A. Nakano, J. Polymer Sci., A-2, 5:125 (1967).

132. A. Nakano and Y. Minoura, J. Appl. Polymer Sci. 15:927 (1971).

133. F. E. Bailey, Jr., and J. V. Koleske, Poly(ethylene oxide), Academic Press, New York, 1976, p. 53.

134. S. A. Dubrovskii, I. V. Kumpanenko, V. M. Goldberg, and K. S. Kazanskii, Polymer Sci. USSR 17 (12):3141 (1975); C.A. 84: 74750 (1976).

135. D. R. Beech and C. Booth, J. Polymer Sci. 46:5127 (1960).

136. S. Nishimoto, B. Ohtani, H. Shirai, S. Adzuma, and T. Tsutomu, Polymer Comm. 26:292 (1985).

137. M. Syczewski, *Przem. Chem.* <u>67</u>:370 (1988); C.A. <u>109</u>:
      190,897 (1988).
138. Yu. A. Mikheev and L. N. Guseva, <u>Khim. Fiz.</u> <u>6</u>:1259
      (1987).
139. E. Calahorra, M. Cortazar, and G. M. Guzman, <u>J.
      Polymer Sci., Polymer Ltr. Ed.</u> <u>23</u>:257 (1985).
140. R. B. Jones, C. J. Murphy, L. H. Sperline, M. Farber,
      S. P. Harris, and G. E. Manser, <u>J. Appl. Polymer Sci.</u>
      <u>30</u>:95 (1985).
141. J. J. Villenave, C. Filliatre, R. Jaouhari, and M.
      Baratchart, <u>Eur. Polymer J.</u> <u>19</u>:575 (1983).
142. L. E. St. Pierre and C. C. Price, <u>J. Am. Chem. Soc.</u>
      <u>78</u>:3432 (1956).
143. R. A. Newton, in <u>Kirk-Othmer Encyclopedia of Chemical
      Technology</u>, Third Ed., Vol. 18, John Wiley & Sons, New
      York, 1979, p. 950.
144. Y. A. Mikheev, L. N. Guseva, L. E. Mikheeva, and Y.
      D. Toptygin, <u>Vysokomol. Soedin., Ser. A</u> <u>31</u> (5):996
      (1989).

6

Physical Properties of Poly(alkylene oxide)s

I. SOLUBILITY, POLY(ETHYLENE OXIDE) (1,2)

Poly(ethylene oxide) is usually considered to be soluble in water in all proportions. However, it has been reported recently that over the temperature range of 20°C to 70°C in very dilute aqueous solution (0.02 to 0.06 percent), an unknown phase exists (3). When this phase is concentrated, it appears to be composed of fibrillar, colloid-sized particles that behave as crystals.

The polymer is also soluble in organic solvents such as acetonitrile, anisole, dichloroethylene, and chloroform. While poly(ethylene oxide) is often thought of as one of the most hydrophilic polymers, it can be quantitatively extracted from water with chloroform. This factor was used in the development of a technique for one of the first molecular weight fractionations of poly(ethylene oxide) (4). This technique employed a liquid-liquid extraction of aqueous solutions of the polymer with a series of chloroform and ethylene dichloride mixtures. The lowest molecular weight molecules were first extracted by solution mixtures containing high ratios of ethylene dichloride to chloroform. This ratio decreased for succeeding fractions.

At room temperature, aromatic solvents such as benzene and toluene are poor solvents for the polymer; but, as the temperature is increased, poly(ethylene oxide) becomes readily soluble. The general solubility characteristics of poly(ethylene oxide) have been summarized by Bailey and Powell (5) and can be found elsewhere (1, 2). Galin (6) studied the solubility interactions with gas-liquid chromatography using solvents that ranged from low-polarity aliphatic hydrocarbons to highly polar trifluoroethanol and determined a value for the solubility parameter at 70°C of 19.64 ± 0.02 $(J/cm^3)^{0.5}$ or 9.60 ± 0.01 $(cal/cm^3)^{0.5}$ (Table 1). Extrapolating these elevated-temperature values for the solubility parameter to room temperature, Galin determined a value at 25°C of 10.3 $(cal/cm^3)^{0.5}$.

Aqueous solutions containing 20 percent poly(ethylene oxide) with molecular weights of several hundred thousand and more are elastic gels. At higher concentrations, water acts as a plasticizer for the polymer, and hard, tough plastic materials result (7). As noted, poly(ethylene oxide), though soluble in water in all proportions at ambient temperatures, has an inverse solubility-temperature relationship. The lower consolute temperature depends on both polymer concentration and polymer molecular weight. In Figures 1 and 2, the lower consolute temperature is shown as a function of concentration at one molecular weight and as a function of molecular weight (or intrinsic viscosity) at one concentration (7,8).

The effect of pH on the lower consolute temperature in water is shown in Figure 3. Very high hydroxide concentrations strongly decrease the solubility of the polymer, while high hydrogen-ion concentrations raise the temperature, indicating that oxonium ions form as proposed by Rosch (9).

Light-scattering studies by Kambe and Honda (10) using both dynamic and static techniques have measured the diffusion of poly(ethylene oxide) in water. In this work, the critical concentration (C_p^*), which marks the transition from dilute to semidilute solution behavior, was determined as a function of molecular weight (Table 2). Above this critical concentration, diffusion of polymer depends on concentration and is independent of molecular weight. Below this critical concentration, diffusion depends on molecular weight and is independent of concentration. From these data, the ratio of the hard-sphere radius to the hydrodynamic radius of the polymer and, hence, the second virial coefficient, can be calculated. A value of 9.25×10^{-4} m^3/kg^2mol was calculated for the second virial coef-

TABLE 1   Solubility Parameter of Poly(ethylene oxide) as a Function of Temperature (From Ref. 6.)

Temperature, °C	Solubility parameter	
	$(J/cm^3)^{0.5}$	$(cal/cm^3)^{0.5}$
25	—	10.3*
70	19.64 ± 0.21	9.60 ± 0.01
100	18.66 ± 0.33	9.12 ± 0.16
120	18.00 ± 0.21	8.80 ± 0.10
150	17.20 ± 0.21	8.41 ± 0.10

*Extrapolated value

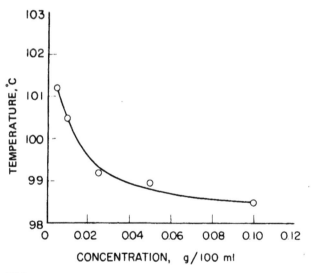

CONCENTRATION, g/100 ml

FIG. 1   Lower consolute temperature of poly(ethylene oxide) in water as a function of concentration, $M_W = 2 \times 10^6$, intrinsic viscosity = 10 dl/g.   (From Ref. 7.)   (Reproduced by permission of the American Chemical Society, Washington, D.C.)

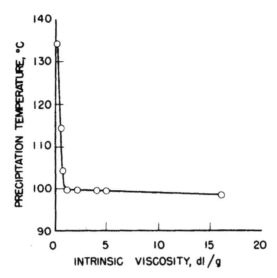

FIG. 2 Lower consolute temperature of poly(ethylene oxide) in water, 1% solution, as a function of intrinsic viscosity (molecular weight). (From Ref. 8.) (Reproduced by permission of John Wiley & Sons, New York.)

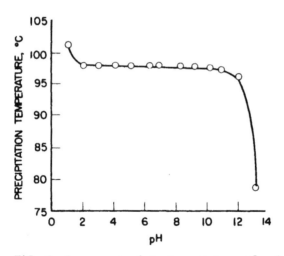

FIG. 3 Lower consolute temperature of poly(ethylene oxide) in aqueous solution as a function of pH. (From Ref. 8.) (Reproduced by permission of John Wiley & Sons, New York.)

TABLE 2   Critical Concentration of Poly(ethylene oxide) in Water at 25°C (From Ref. 10.)

Molecular weight	$C_p^* \times 10^3$, g/ml (dynamic scattering)	$C_p^* \times 10^3$, g/ml (static scattering)
1.77×10^5	8.5	6.0
3.00×10^5	6.9	5.0
6.79×10^5	3.9	4.6
1.19×10^6	3.1	3.9

ficient of poly(ethylene oxide), as compared to a value of 8.3×10^{-4} m^3/kg^2mol independently determined by Roots and Nystrom (11) for a 300,000 molecular weight polymer.

Light-scattering (12) and photoelectron correlation spectroscopy with time-averaged intensity measurements (13) have confirmed that poly(ethylene oxide) molecules aggregate in water solution as temperature is raised from 20°C to 90°C. This aggregation has been ascribed to hydrophobic interactions becoming dominant at higher temperatures.

The precipitation temperature of poly(ethylene oxide) from water solution—that is, the lower consolute temperature—would be due to an increase in activity of the polymer. The Debye-McAulay equation (14),

$$\ln f_{(s)} = (\beta/2kTD_o) \sum_i (n_i e_i^2/b_i)$$

in which β is an empirical constant; $f_{(s)}$ is the fugacity of the neutral molecule solute in water; k, T, and D_o are the Boltzmann constant, Kelvin temperature, and solvent dielectric constant, respectively; $\Sigma\, n_i e_i^2$ is the salt ionic strength (concentration); and b_i is the ionic radius of ion i, predicts a salting out that increases with salt ionic strength and is strongest for salts of small ions.

The effect of a number of salts in lowering the consolute temperature of poly(ethylene oxide) in water is shown in Figure

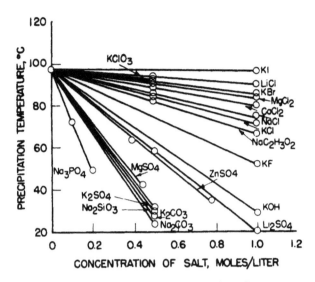

FIG. 4   Effect of salt concentration for various salts on the
lower consolute temperature of poly(ethylene oxide) in aqueous
solution. (From Ref. 8.) (Reproduced by permission of John
Wiley & Sons, New York.)

4.   The lower consolute temperature decreases essentially linearly with salt concentration, but the effectiveness of salts does
not follow the ionic strength principal.   The order of the ions
in effectiveness in salting out the polymer, shown in Table 3,
resembles the Hofmeister series for proteins (15) and is similar
to that found by Schick in the study of the critical micelle concentration for polyoxyethylene nonylphenol solutions (16).

Amu (17) critically reviewed the effect of salts on the solubility of poly(ethylene oxide) in water together with intrinsic
viscosity measurements and Flory theta conditions.   From this
work, Amu estimated the theta temperature of poly(ethylene oxide) in water to be 108.5°C.   Amu presented data in terms of
the Stockmayer-Fixman equation (18),

$$[\eta]/M^{0.5} = K_G + 0.51 \Phi B M^{0.5}$$

where $[\eta]$ is the intrinsic viscosity, Φ is the Flory constant,
and M is the weight average molecular weight.   The temperature dependence of the parameter B is, following Flory (19),

TABLE 3 Approximate Salting-Out Effectiveness of
Ions on an Ionic Strength Basis

For poly(ethylene oxide)	Small neutral molecules
(In order of decreasing effectiveness)	
Anions:	
OH^-	OH^-
F^-	
CO_3^{-2}, SO_4^{-2}	CO_3^{-2}, SO_4^{-2}
$C_2H_3O_2^-$	
Cl^-	Cl^-, $C_2H_3O_2^-$, ClO_3^-
PO_4^{-3}	
Br^-, ClO_4^-	Br^-, I^-
Cations:	
Na^+, K^+	Na^+, K^+
Li^+	
Ca^{+2}, Mg^{+2}	Li^+, Ca^{+2}, Mg^{+2}, Zn^{+2}
Zn^{+2}	
H^+	H^+

$$B = B_0[1 - (\theta/T)]$$

$$B = (-65.7 \times 10^{-27})(1 - 381.5/T)$$

and, hence, the Flory theta temperature has a value of 108.5°C
or 381.7 K. These data also predict the temperature dependence of the exponent a in the Mark-Houwink equation, which

relates intrinsic viscosity to viscosity average or weight average molecular weight.

$$[\eta] = KM^a$$

For poly(ethylene oxide) in water at 30°C, a value of 0.87 was calculated for the exponent a

$$a_{(30°C)} = (552.8/T) - 0.95 = 0.87$$

which can be compared with the following published value (1) for the Mark-Houwink equation at 30°C for the molecular weight range of 10^4 to 10^7.

$$[\eta] = 1.25 \times 10^{-4}M^{0.78}$$

This reference (1) contains constants for intrinsic viscosity-molecular weight relationships that have been determined for a number of molecular weight ranges, solvents, and temperatures.

One interpretation of the inverse solubility-temperature relationship of the poly(ethylene oxide) chain in water solution—that is, the explanation for the lower consolute temperature—has been that there is a hydrophilic-hydrophobic balance that exists in which water molecules are oriented relative to the polyether chain via hydrogen bonding. Indeed, it could be estimated that three water molecules were closely associated with each oxyethylene unit in the chain, wrapping around the polyether chain in a hydrogen-bonded, helical fashion. As the temperature was raised, this sheath of water molecules dissociated, just as the molecules in the bulk water phase tend to loosen up the dipole alignment with increasing temperature. As the associated water molecules are held less tightly and are lost, the polyoxyethylene chains first form aggregates in solution and then precipitate as the lower consolute temperature is reached.

To this explanation, observations must be added that will be elaborated upon in the following sections. From viscosity data, it can be shown that the entropy of dilution of poly(ethylene oxide) in water is negative. The rheology of somewhat more concentrated poly(ethylene oxide) solutions in water is entropy controlled. The cations of salts, such as sodium or potassium, do form complexes with the polyether; in fact, polyethylene glycols and, more particularly, higher molecular weight poly(ethylene oxide)s, behave like weak crown-ethers, holding the cations within the coiled polyether helix. Such salt-poly-(ethylene oxide) complexes do behave like polyelectrolytes in solution. Salts will form complexes that greatly depress the melting point of the usually crystalline polymer, and these polymer-salt compositions, which are amorphous at ambient temperature, can have very high ionic electrical conductivity (1).

The explanation of the lower consolute temperature for the polyoxyethylene chain in water solution, and the salting-out behavior, based on earlier considerations outlined above, is likely incomplete. The earlier treatment pictured the polyoxyethylene chain, above some minimum molecular weight of about 1500, as a highly expanded random coil and the collapse of that coil, particularly in salt solutions, as leading to theta conditions of the Flory type. A critique of this interpretation was presented by Amu (17). The question is whether or not the random coil in water-salt solutions is simply collapsed, in the Flory sense, by a salting-out phenomenon that could be described by the reasoning of the Debye-McAulay equation.

The answer probably is that ion associations give rise to a rather different and more complex phenomenon than only dehydration of the polyoxyethylene chain in water solution. In addition, ion-clustering effects in the poly(ethylene oxide) molecule in water-salt solutions probably occur even in dilute solutions.

The crown-ether-like associations (20) of specific cations (Table 3 and Figure 4) change the hydrodynamic conformation of the expanded random coil to a considerably more anisotropic conformation, noting that there are both ether-bound cations and associated anions present. The result should be an elongated molecule, rather than a slightly oblate, spheroid-shaped polymer molecule. Consequently, the conformation would have a larger excluded volume. The entropy change of such a polymer molecule in passing from solution to a separate phase may be small. In the case of poly(ethylene oxide) in aqueous salt solutions, following Flory's arrguments (21), transfer of salt complexed with polyoxyethylene in water solution to a water-salt solution while the polymer separates as a separate phase, which may or may not also contain water, can result in a net increase in entropy. The phase separation of poly(ethylene oxide) from water or water-salt solution is entropy-driven.

It is possible to go back to the first observations made in the this section about the solubility of poly(ethylene oxide).

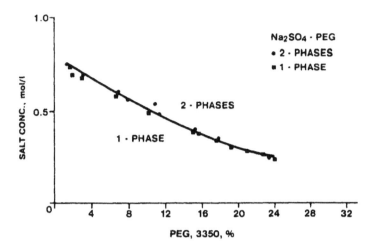

FIG. 5 Phase diagram for the formation of aqueous two-phase systems in polyethylene glycol-Na₂SO₄ solutions. Phase formation occurs above the curve. (From Ref. 21.) (Reproduced by permission of the American Chemical Society, Washington, D.C.)

While poly(ethylene oxide) is often considered to be one of the most hydrophilic of existing polymers, it can be extracted from water solution by chloroform. This transfer of the polymer from water to an organic solvent is also entropy-driven, with the polymer assuming a random coil conformation in chloroform and freeing bound or complexed water in the water-poly(ethylene oxide) solution.

This model for the water solubility-temperature relation of the polyoxyethylene chain has been explored in detail by Ananthapadmanabhan and Goddard with regard to biphase formation from water solution of polyethylene glycols in the molecular weight range of about 1500 to 20,000 (21). In this work, the effectiveness of various salts in forming aqueous two-phase systems with polyethylene glycols was determined (Figures 5, 6, and 7). For each salt, the region below the line in the figures represents homogeneous solution, while in the region above the line, two phases are present.

From Figure 7, it appears that salts of higher valence anions form two-phase systems—that is, salt out the polyethylene glycol—at lower concentrations than do those of monovalent anions (see also Figure 4). Salts with monovalent cations, how-

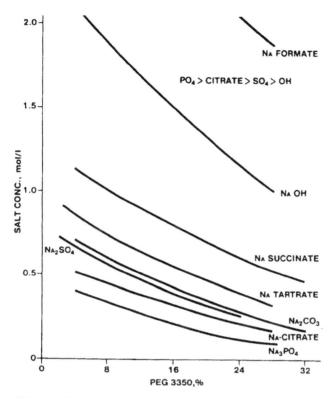

FIG. 6 Phase diagrams for the formation of aqueous two-phase systems in polyethylene glycol-sodium salt solutions. Phase formation occurs above the curve. (From Ref. 21.) (Reproduced by permission of the American Chemical Society, Washington, D.C.)

ever, appear more effective in lowering the consolute temperature than do polyvalent cations. Ananthapadmanabhan and Goddard (21) note that the effectiveness of anions in salting out polyethylene glycols follows the lyotropic series (23) (Table 4).

Salts that are most effective in lowering the consolute temperature of polyoxyethylene in water have cations that tend to form strong crown-ether-like complexes with the polyoxyethylene and anions that are of low lyotropic number. Salts that tend to form cation complexes with polyoxyethylene and have anions with high lyotropic number show high ionic conductivity (24, 25) in the solid phase with poly(ethylene oxide).

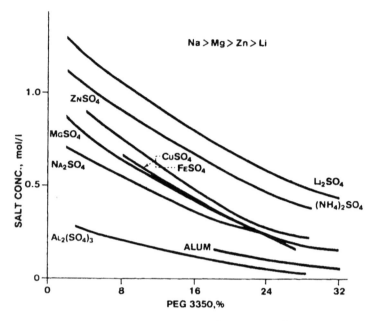

FIG. 7 Phase diagrams for the formation of aqueous two-phase systems in polyethylene glycol-metallic sulfate solutions. Phase formation occurs above the curve. (From Ref. 21.) (Reproduced by permission of the American Chemical Society, Washington, D.C.)

Generally, enthalpic contributions favor the dilution of the salt and poly(ethylene oxide) to give a single phase. The entropic factors can be calculated using the following expression for each phase:

$$\Delta S = -R \sum x_i \ln \phi$$

where x_i = mole fraction of species i, ϕ = volume fraction of species i, and R is the gas constant. For phase separation to overcome the enthalpic factors, the entropy change for formation of two phases must be

$$\Delta S_{two\ phases} > \Delta S_{one\ phase}$$

TABLE 4   Lyotropic Series

Anion	Lyotropic number	Formation of biphase with polyethylene glycol
SO_4^{-2}	2.0	Yes
PO_4^{-3}	3.2	Yes
F^-	4.8	Yes
OH^-	5.8	Yes
Cl^-	10.0	No
Br^-	11.3	No
NO_3^-	11.6	No
I^-	12.5	No
SCN^-	13.3	No

or the sum of the volume factors must be less for the two phases than for the single phase. This is a circumstance that fits the rationale that the polymer-salt complex excluded volume in the single-phase conformation at the consolute temperature is greater than the sum of the salt in the salt-rich phase and the polymer in the polymer-rich phase after phase separation.

Interestingly, the effectiveness of salts in lowering the consolute temperature of polyoxyethylene in water exactly parallels the effectiveness of salts in lowering the cloud point of ethoxylated nonionic surfactants (16, 26).

The electrical properties of the alkali metal salt complexes are of interest as solid electrodes for batteries (27, 27a). There is particular interest in alkali metal-thiocyanate salt complexes; this interest is related to the anion having a high lyotropic number and a correspondingly high conductivity (see Table 4). The complete phase diagram has been published for the potassium thiocyanate/ and sodium/thiocyanate/polyethylene glycol system (28), and various thermodynamic properties have been determined from viscosity, cloud point, osmotic pressure, and gel permeation chromatography measurements (29). Boils-Boissier determined the effect of high-molecular-weight poly-(ethylene oxide), 300,000, which is considerably less crystalline (69 percent) than a 4000-molecular-weight polyethylene glycol (96 percent) used in a comparative study, on the properties of the potassium thiocyanate/polyethylene oxide complex (30). Use of high-molecular-weight poly(ethylene oxide) did not affect the nature of the equilibrium that exists during complexation, but it did cause a 6°C increase in the eutectic melting temperature and in the melting of the well-defined complex. The latter increase was thought to be related to an increase in the thickness of the crystallites obtained from the high-molecular-weight system. A recent study (31) was involved in an investigation of the effect of cation nature and size, as well as the nature of the solvent used for complexation of polyethylene glycol. When acetonitrile was used as the solvent and the anion was SCN^-, the binding capacity of the cations decreased in the order $Cs^+ > K^+ > Na^+ > NH_4^+ > Li^+$. When barium and strontium were used to complex poly(ethylene oxide) in methanol, the anion binding constant was very dependent on the type anion and decreased in the order $ClO_4^- > SCN^- > Br^- > Cl^-$ (32).

Poly(propylene oxide) polyols also form complexes with potassium and sodium thiocyanate. Teeters and coworkers (33) used glass transition temperatures from differential scanning calorimetry (DSC) measurements to investigate the complexation of a 3000-molecular-weight triol with these alkali metal salts. The results indicated that at least 10 ether atoms per cation were required for formation of stable complexes.

II. CRYSTALLIZATION BEHAVIOR

A. Poly(ethylene oxide)

Poly(ethylene oxide) has been investigated in both a single-crystal form and in normal or commercial, partially crystalline form. Single crystals have been examined by X-ray analysis and have been shown to be flat platelets with the c-axis of the unit cell oriented normal to the basal plane (34). Electron microscopy has shown bulk crystallized polymer to have a lamellar structure (35, 36), and single-crystal lamellae have been grown from dilute solution (37, 38).

The degree of crystallinity of commercial products is highly dependent on molecular weight, with the low-molecular-weight

(>~1000) polyethylene glycols having high degrees of crystallinity (>90 percent) and the high-molecular-weight products
having only moderate degrees of crystallinity (~50 percent).
Products with molecular weights of less than about 800 are liquids under normal ambient conditions. The crystalline poly-
(ethylene oxide) chain contains seven structural units,
$-CH_2CH_2O-$, in two helical turns for each identity period.
There are four molecular chains in the monoclinic crystallographic unit cell, which has a = 7.96A, b = 13.11A, and c = 19.39A
and a characteristic angle of 124° 48'. The chain array has di-
hedral symmetry with twofold axes. One axis passes through
the ether-oxygen atoms and the other bisects the carbon-to-
carbon bond (39—49). Internal rotations about the $-O-CH_2-$,
$-CH_2CH_2-$, and $-CH_2-O-$ bonds have the conformational as-
signments trans, gauche, and trans, respectively (45—46, 49—
50).

Crystallization kinetics have been studied in bulk and in
dilute solution by a number of investigators (51—57). In general, these investigations were based on either the Avrami
theory (58) or the Hoffman-Lauritzen theory of polymer crystallization (59—61).

From the melt, poly(ethylene oxide) crystallizes in two
stages. First, nucleation takes place; this is then followed by
growth of the nuclei to form spherulitic structures, which can
be of macroscopic size, and lamellae. Different growth rates
correspond to the relative rates of nucleation and growth of a
crystalline layer at the growth front. Godovsky (62) and coworkers used calorimetric, dilatometric, and hot-stage microscopy to study polyethylene glycols with molecular weights of
400 to 20,000. From the Flory relationship for melting (19),

$$(1/T_m^0) = (1/T_m) - 2R/H_f X$$

where T_m^0 is the equilibrium melting point for an infinite molecular weight polymer, T_m is the observed melting point, R is
the gas constant, H_f is the molar heat of fusion per chain unit,
and X is the degree of polymerization, they determined a value
of 70°C for T_m. Recently (63), a value of the equilibrium
melting temperature was determined with a 100,000-molecular-
weight poly(ethylene oxide) using the Hoffman-Weeks extrapola-
tion technique (64, 65), and a value of 69°C was found. The
nucleation theory of chain folding has been extended to predict

the degree of undercooling at which a chain of given length
will display a given number of folds per molecule (66).

Godovsky et al. found Avrami exponent values of 2 ± 0.6
for the polyethylene glycols studied. The Avrami equation,

$$(1 - X_c) = \exp(-kt^n)$$

where X_c is the fraction crystallized at time t, k is the crystallization-rate constant and is dependent on nucleation and growth
rates, and n is related to growth geometry and type of nucleation, should yield a value of $n = 3$ for poly(ethylene oxide)
(67—70). Beech and coworkers (71, 72) investigated the behavior of fractionated poly(ethylene oxide) and concluded that this
deviation was due to different secondary processes that were
molecular-weight-dependent. As crystallization proceeds in
polymers with a molecular weight of <6000 or in those with a
broad molecular weight distribution, there is a fractionation by
molecular weight and an attenuation in crystallization rate.
When high polymer is involved, the secondary process involves
a perfection of the folded-chain crystals through lamellar thick-
ening.

The fusion of fractions of poly(ethylene oxide) over the
molecular weight range of 3500 to 5,000,000 has been studied by
thermal analysis (73—76). Complete segregation of high-molecular-weight polymer from polymers up to 20,000 molecular weight
was possible. The segregation is controlled by molecular nucleation rather than by the melting/crystallization equilibrium.
As crystallization temperature was increased, three regions of
growth were found. The first region involved formation of
mixed crystals from the high- and low-molecular-weight polymers. This was followed with a region of sequential crystallization and partial segregation, and finally a region of total segregation. The morphology of crystalline polyoxyethylene is
more complex than the integer-fold concept in which all segments of a molecule are the same length with the chain ends located at the lamellar surfaces. It has been pointed out that
both mixed-integer and fractional-integer chain folding exists
in the lamellae (75).

The self-seeded crystallization of high-molecular-weight
poly(ethylene oxide) from dilute toluene solutions by static and
dynamic light-scattering techniques indicated that the radii of
the crystals grow in a linear manner (77). The rate constant,

which is obtained from a plot of crystal radius versus time,
depends on temperature and molecular weight. The growth of
a 5,000,000-molecular-weight polymer was about 35 times faster
than that of a 100,000-molecular-weight polymer. A change in
temperature from 25°C to 15°C resulted in over a 500-fold increase in the rate constant for the 100,000-molecular-weight
polymer. This suggests that diffusion of the polymer chains to
the crystal surface is not the rate-determining step. Final
crystal size was greater as the crystallization temperature decreased. This was related to polymer fractionation during crystallization, wherein lower molecular weight chains have a lower
crystallization temperature than do high-molecular-weight chains.

Gas-solid chromatography has been used to determine the
degree of crystallinity of polyoxyethylene (78, 79). Galin (79)
points out that it is necessary to take adsorption at the probe
if reliable values are to be obtained.

The effect of poly(methyl methacrylate), PMMA, on the
crystallization kinetics of poly(ethylene oxide) has been investigated using the Avrami equation to analyze the results (80).
The crystallization-rate constant, k, decreased as the concentration of PMMA increased. This and other results indicated
that, in the blends, crystallization proceeds by a predetermined
nucleation and this is followed with a two-dimensional growth.
There has been evidence of melt compatibility for these two
polymers (81—84); see Section V. Crystallization behavior of
blends of poly(ethylene oxide) with poly(propylene oxide) (85)
and with poly(vinyl acetate) (83) have been studied, as well as
star and block copolymers of ethylene oxide and styrene (86).

Nuclear magnetic resonance was used to investigate the effect of gamma radiation on the morphology of semicrystalline
poly(ethylene oxide). The changes were related to crosslinking
and to degradation (see Chapter 5). The growth that took
place in a down-field peak in the carbon-13 spectra at high radiation dosages was related to rearrangements in the crystalline
domains (87).

Macrospherulites of polyethylene glycols that have diameters
of up to about 2 cm (Figure 8) have been reported (88). When
viewed between crossed Polaroids®, the characteristic Maltese
cross patterns and spherulitic banding that are seen in micro-
scopic spherulites are apparent in these spherulites and can be
seen with the naked eye (Figure 9).

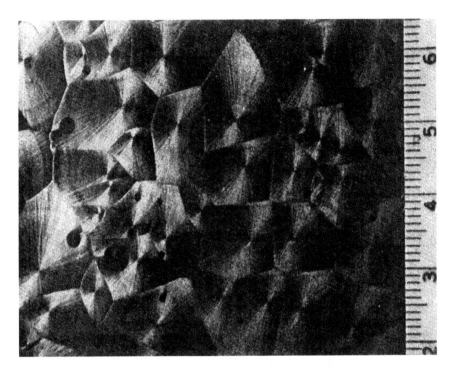

FIG. 8 Macrospherulites (spherulites that are observable without a magnifying device) of a 1540-molecular-weight polyethylene
glycol. Magnification is given by the centimeter scale on the
photograph.

B. Poly(propylene oxide)

Poly(propylene oxide) crystallizes with an orthorhombic unit
cell that has the dimensions a = 10.40A, b = 4.67A, and c =
7.16A and that contains two chains in a bowed zigzag conformation (89). The methyl group is between the oxygen and methylene groups of the adjacent molecular chain. It was felt that
a planar zigzag conformation would be under a great deal of
strain and that bowing of the chains was due to the proximity
of a methyl group to the methylene group of the neighboring
unit on the chain three atoms away. Other studies reported

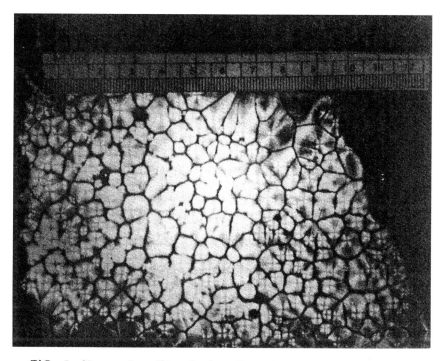

FIG. 9 Macrospherulites (spherulites that are observable without a magnifying device) of a 1540-molecular-weight polyethylene
glycol taken through crossed Polaroids. Magnification is given
by the centimeter scale on the photograph.

similar but slightly different unit cell dimension (90−92).
Melting point of the polymer was reported as 72−74°C (57),
and values of the glass transition temperature were reported to
be between −73°C and −78°C when determined by dilatometry
(93−95).

Allen, Booth, and Jones (96) investigated the intrinsic viscosity-molecular weight behavior of amorphous and crystalline
fractions, as well as whole polymers of poly(propylene oxide),
and found that their combined behavior could be described by
the following Mark-Houwink equations.

$$[\eta] = 1.12 \times 10^{-4}(M_w)^{0.77}, \text{ benzene at } 25°C$$

$$[\eta] = 1.29 \times 10^{-4}(M_w)^{0.75}, \text{ toluene at } 25°C$$

Molecular weights were determined by light scattering.

C.   Polyepichlorohydrin

Isotactic polyepichlorohydrin crystallizes with an orthorhombic unit cell that has the dimensions a = 12.6A, b = 4.90A, and c = 7.03A and a zigzag conformation that differs from planar by 15 degrees (97, 98).   There are four monomeric units in the unit cell.   The chloromethyl side group lies on a plane essentially perpendicular to the main chain axis.   Calculated density is 1.47 g/cm^3.

Broad-line NMR over the temperature range of −170°C to 40°C was used to investigate the molecular mobility and phase structure of partially crystalline, isotactic polyepichlorohydrin (99).   The onset for rotation of the chloromethyl side group occurred at −130°C.   The onset of cooperative long-chain segment rotational and translational movements was −30°C, in agreement with the DSC-determined value of −27.5°C (100). Although X-ray diffraction indicated that the polymer was 38 percent crystalline, NMR indicated that 70 percent of the chains remained rigid at the glass transition temperature, which was felt to be related to isotactic chain conformation in contrast with amorphous chain conformation (101, 102).

III.   ASSOCIATION COMPLEXES (103)

A.   Poly(ethylene oxide)

The ether oxygen atoms in poly(ethylene oxide) chains have a strong hydrogen-bonding affinity that results in the formation of association or interpolymer complexes when the polymer chain interacts with polymeric acids such as poly(acrylic acid) and poly(methacrylic acid) in solution (104) and with compounds such as urea and thiourea in bulk (105−107).   The subject of association complexes has been reviewed with a particular emphasis on the polyoxyethylene complexes (108) and on polyoxyethylene-based surfactant complexes (109).

When finely divided urea or thiourea is suspended in a dilute (1–3 percent) benzene solution of polyoxyethylene, a complex forms. The polymer can be quantitatively removed from
solution by this technique. The complex forms in a ratio of
about two molecules of the suspended compound for each ethylene oxide unit. This closely corresponds to the expected ratio
from a helical arrangement about any oxyethylene unit in a
zigzag conformation. Tadokoro and coworkers (110) investigated
the urea-polyoxyethylene complex obtained by immersing an
oriented film of the polymer into a methanolic solution of urea.
From X-ray analysis and formation/destruction/formation of the
complex, they found that the polyoxyethylene molecule in the
complex does not have a planar zigzag conformation. Instead,
the complex has a helical structure similar to that found for the
polymer alone in the crystalline state. Wright (111) has reviewed the crystallographic and morphological structure of the
semicrystalline alkali metal salts with poly(ethylene oxide)s.
Included in the review are recently prepared poly(ethylene oxide) complexes that have a mixed electronic-ionic mechanism of
conductivity. Lattice energy for the sodium iodide complex has
been calculated from the lattice energies of pure polymer and
pure salt (112).

Complexes similar to these have been formed from a variety
of organic compounds. Blumburg et al. (113) formed a mercuric
chloride complex by suspending a film of the polymer in a saturated solution of $HgCl_2$ in dry ethyl ether. The complex had
a ratio of one mercuric chloride molecule per four oxyethylene
units and was rigid, brittle, and insoluble in water. The complex was also studied by Tadakoro et al. (114), who found that
different complexes formed depending on immersion time. Complexes with differing properties have also been formed from
bentonite (115), cadmium chloride (112), resorcinol (116), and
sodium dodecylsulfate (117).

Smith et al. (104) discovered that when aqueous solutions
of poly(ethylene oxide) (POE) and poly(acrylic acid) (PAA)
were mixed in approximately equal proportions, a precipitate
immediately formed. The precipitate could be molded to form
clear, water-insoluble films and was found to be an intermolecular association complex of the two polymers. This complex is
affected by hydrogen-ion concentration. Below a pH of 3.8,
the complex precipitates. Precipitation also occurs above a pH
of 12.3, but it is thought that only POE precipitates in this
high-pH region (118). If a water-soluble consolute such as

dioxane or acetone is added to the aqueous solution, complex precipitation is inhibited at low pH values and higher hydrogen-ion concentrations are required to precipitate the complex.

These very interesting complexes of high-molecular-weight polymers have been the subject of many papers, and interest remains high today. In general, two types of complex formation can occur. These are described as stepwise complexation and one-to-one complexation (119), and the POE/PAA is of the latter type. A complex formed between poly(vinyl pyrrolidone) and a phenolic in an acetone/methanol mixture forms with stepwise kinetics, whereas with poly(ethylene oxide) and the phenolic in the same liquid system, it forms by one-to-one complexation. Formation of the latter complex can be followed with infrared spectroscopy because the ether oxygen absorption shifts from 1095 cm^{-1} to 1070 cm^{-1} and the phenolic $-C-O$-absorption shifts from 1225 cm^{-1} to 1250 cm^{-1}.

Not all polyoxyethylene association complexes form by one-to-one complexation. The polyoxyethylene/p-hydroxy-benzoic acid-formaldehyde copolymer complex in methanol has been reported to form by stepwise complexation (120, 121). Uncertainty regarding the type of complexation may be due to changes in molecular weight distribution via partial fractionation that occurs during the complexation process (122).

Most studies of interpolymer or association complexes have been carried out in either aqueous media or in organic solvents. Chatterjee et al. (123) investigated complex formation between POE and PAA in tetrahydrofuran/water mixtures and found behavior different from that reported for water or organic solvent alone. Although several factors appeared to be in effect at different solvent ratios, including preferential solvation of the polymer, probable changes in conformation of the PAA, and disordering of water molecules bound to the POE, the complex was always soluble and always was found in the one-to-one unit mole ratio. A new viscometric method has been used to investigate the formation and structure of POE/PAA complexes (124). Molecular weight, degree of neutralization, and ionic force variable were examined.

An interesting study by Morawetz and coworkers (125) involved the kinetics of POE/PAA complexation. The kinetics were studied by means of fluorimetry using a tagged PAA. This study indicated that the equilibrium reached was remarkably insensitive to POE molecular weight over the range of 8000 to 80,000. It has also been found that the change in fluores-

cence of dansyl-labeled PAA allows the kinetics of complexation
and complex decomposition to be followed (126). When Bednar
and coworkers (127) studied complex formation between POE
and PAA by following the changes of emission of dansyl-PAA,
it was observed that addition of a 24,000-molecular-weight POE
to the labeled PAA resulted in up to an eightfold increase of
fluorescence with a blue shift of the maximum. When much
higher molecular weight POE was used, relatively small increases in fluorescence with a red shift of the maximum were
observed. This latter factor suggested that the dansyl-PAA
was stretched out and only in contact with POE at widely separated regions. Addition of 24,000-molecular-weight POE to the
latter complex resulted in changes in emission similar to those
obtained in the absence of the high-molecular-weight POEs.
Also, removal of the dansyl-PAA with PAA led to sharp decreases in fluorescence; but, as time passed, it led to an increase in fluorescence. This indicated that the labeled PAA is
freed in a stretched state before contracting to its equilibrium
conformation. In another study, pyrene groups were attached
to the ends of polyethylene glycol molecules by direct esterification with 1-pyrenebutyric acid (128). The end-labeled molecules allowed excimer fluorescence to be used as a probe for
the complexation of different molecular weight glycols and poly-
(acrylic acid)s of very widely varied molecular weight ($M \sim 2000$
to 900,000). Addition of PAA to the glycols resulted in sup-
pressed glycol intramolecular mobility, and the concentration of
glycol increased in the locality of PAA due to hydrogen bond
interaction. Acanaphthylene has been used to label PAA and
demonstrate that POE/PAA complexation resulted in a sterically
restricted, tightly packed conformation (129). As usual, complexation was maximized at a 1:1 ratio of monomer moieties.
When less than 10 percent of the carboxyl groups were neutralized, the polarization of the PAA probe was restored to its un-
complexed state. Oyama, Tang, and Frank (130) have reviewed
the use of X-ray fluorescence for complexation studies of py-
rene-containing polyoxyethylene glycol/poly(acrylic acid) systems formed in water.

   Complex-forming POE/PAA interpenetrating polymer networks have been prepared by matrix polymerization of acrylic
acid and ethylene diacrylate (a crosslinking agent) with a pre-
crosslinked POE that was formed from hydroxyl-terminated POE
and 2-isocyanatoethyl-2,6-diisocyanatohexanoate (131). At low
degrees of POE crosslinking, dynamic mechanical properties in-

dicated that only the 1:1 POE/PAA acrylic acid complex existed. As the degree of POE crosslinking was increased, three phases existed: a POE-rich phase, a 1:1 complex phase, and a PAA-rich phase. At high degrees of POE crosslinking, only two phases were present—a POE-rich and a PAA-rich phase—indicating that complex formation was not possible in the tightly crosslinked networks. A potential use for these complexed networks is as a special function membrane (132) in which the membrane acts as a chemical valve whose permeability can be controlled by altering the chemical environment through ionic buffer strength and pH. Flux through the membrane and size of solute through the membrane can be changed in this manner. The pH-dependent mechanochemical reaction involves at least two separate processes (133). One process is related to the interpenetrating networks' viscoelastic creep recovery, and the other process is related to the reversible association and disassociation of the molecules in the complex.

The complexation of cobalt-60 radiation-crosslinked POE and of linear POE with molybdenum-VI salts indicated that, in the linear form, complexation is extremely weak and of low stability—40 to 80 times weaker than that with sodium salts (134). This was related to vacant d-orbitals in molybdenum being involved in the interaction process. This resulted in greater polarizability of the cations and steric restriction in coordination with the ether oxygen atoms of the polymer chains. However, when the salts were complexed with the crosslinked POE, stability constants two orders of magnitude higher than those obtained with linear POE were found. This was interpreted as being related to intermolecular rather than intramolecular complexation. Stoichiometry for the crosslinked POE system was 4.5 ethylene oxide units per molybdenum cation.

The polysaccharides dextran, dextrin, and inulin form association complexes with polyoxyethylene (135, 136). Such complexation is important to the interaction of synthetic polymers with biological cell membranes, since complex formation has the potential of increasing membrane fluidity or inducing molecular aggregation.

Complexes also form between POE and poly(methacrylic acid), PMAA, with a stoichiometric ratio of three ethylene oxide units for each methacrylic acid unit when atactic PMAA is used. There is evidence that stoichiometry varies when isotactic PMAA is used. It appears that the association complexes that form between POE and polymeric acids involve hydrogen bonding be-

tween carboxylic and ether groups and a hydrophobic bond interaction. Also, the stoichiometry of the complex formed is
probably affected by the steric structure of the polymeric acid.
It is interesting to point out that when an acrylic acid copolymer containing 9 mole percent acrylamide is mixed with POE, a
complex similar to those of PAA and POE forms, but the
acrylamide moieties do not complex with the POE units (127).
This suggests that the presence of uninterrupted sequences of
interacting groups is not important to POE/PAA complex formation; but, as described above, certain limits do exist in crosslinked systems (131). A pH-reversible complex between poly-
oxyethylene/polystyrene block copolymer and PMAA has been
investigated (137), as well as between polyethylene glycol and
a random copolymer of methacrylic acid and methyl methacrylate
(138). The noncomplexing moieties, as methyl methacrylate,
deleteriously alter the perfection of complexation. Calorimetric
and Fourier transform infrared (FTIR) spectroscopic analyses
were used to investigate the hydrogen bonding that exists in
solid-state complexes formed from PMMA and polyoxyethylene
(139). The results were further, new evidence that the complexes form in a one-to-one mole ratio.

Oligomeric oxyethylene methacrylate has been synthesized,
polymerized, characterized, and used to design complexes that
yield high-conductivity, solid, polymeric electrolytes (140).
These complexes have cationic, single-ion conduction that is
dependent on electrolyte content, dissociation energy, and degree of polymer segmental motion for the segments surrounding
ions in the matrix.

Studies that were slanted toward biological applications,
such as muscle contraction, involved development of a mechanochemical energy-conversion system from a PMAA/polyethylene
glycol complex (141, 142). The system functioned through a
thermally reversible, complexation/dissociation mechanism. When
loaded to 100 times its weight and heated from 10°C to 60°C,
the complex in membrane form contracted over 90 percent.
This contraction produced work equal to 5×10^{-3} cal/g of complex, which is about equal to that produced by natural muscle.
Other features of the study involved determination of complex
stability and the ratio of interacted groups to the total number
of groups capable of interacting in both aqueous and water-
ethanol media. Complexation and aggregation of PMAA and
polyethylene glycol complexes at different pHs and temperatures
were investigated by means of turbidity measurements, laser

light scattering, and scanning electron microscopy (143). The complex that is initially formed in aqueous media aggregates through desolvation and hydrophobic interaction into almost spherical agglomerates that grow as a function of time. Decreasing pH over the range of 2 to 7 and increasing temperature over the range of 20°C to 50°C increased aggregation rate.

Association of certain electrolytes has been found for the POE molecule in solution. For example, anhydrous methanol is a nonsolvent for high-molecular-weight POE at room temperature. However, if a small amount of certain metal halides is added, the methanolic solution becomes a good solvent. The nature of this salting-in phenomenon has not been resolved, but it may be due to ion binding, which prevents coagulation. It has been found that as little as 0.5 percent potassoum iodide renders anhydrous methanol a good solvent for POE (108). Plots of viscosity as a function of KI concentration in methanol for POE with a molecular weight of 500,000 have an appearance that is reminiscent of the electroviscous phenenomon associated with polyelectrolytes. The viscosity increases at low ion concentration were interpreted in terms of the nonionic POE becoming a polyelectrolyte by ion binding in certain nonaqueous solvents. The binding of sodium and potassium iodides and thiocyanates to polyethylene glycols in methanol has been studied by means of conductivity measurements (144). The results confirmed the adsorption of cations to specific sites, forming a charged polymer. Solid- and solution-state properties of polyoxyethylene/poly(acrylic acid) complexes formed in methanol and in water with and without calcium sulfate present have been investigated by a wide variety of techniques (145). Complexation was clearly evident in all cases.

Liu (146) used NMR studies to confirm ion associations of the POE chain in solution. Evidence was first found for the binding of potassium iodide in methanol and hydrochloric acid in aqueous solution to poly(ethylene oxide), with the cation being the principal interacting species. These investigations led to conjecture that the mechanism of interaction was akin to that of the crown-ether complexes (147, 148). Sodium-23 NMR measurements have verified the association of the sodium ion with the ether oxygen of the ethylene oxide unit in solvents such as acetonitrile (149). Florin (150) measured NMR relaxation rates for lithium-7, sodium-23, cesium-133, chlorine-35, and bromine-81 complexes and found similar results; i.e., there was asymmetric hydration of the ions induced by polyoxyethylene,

and there were direct cation/ether-oxygen interactions of the
same type, though weaker, as those found in metal complexes
with crown-ethers. A double helical model has been proposed
for crystalline complexes of alkali metal ions with poly(ethylene
oxide) (151). Ion transport and conductivity have been measured by various investigators (152—160). It has been demonstrated that transport of the lithium ion is markedly influenced
by salt concentration and the type of counterion used. The
viscoelastic (161), solution (162), and phase-diagram (163)
characteristics of the aqueous salt complexes have been described.

Graham and coworkers used water-vapor pressure and differential scanning calorimetry measurements to investigate the
interaction of linear polyethylene glycols with water (164).
They found definite evidence for the existence of stable, crystalline, water/polyethylene glycol complexes. The complexes
form only when the molecular weight is greater than 440; but,
when formed, are stable up to temperatures of at least 24.5°C.
The complexes are usually formed from three molecules of water
per ether oxygen, but they also exist at lower ratios. This
complexation of water with the ether oxygen of oxyethylene
units has been shown to be the reason for water solubility of
ethylene oxide/propylene oxide block copolymers (165). Another study (166) has shown that three phases exist in the
water/polyethylene glycol/glycerol system at elevated temperatures of 383 K to 415 K.

Triblock copolymers of poly(methyl methacrylate) at terminal ends with side chains containing an average of eight oxyethylene units each and a middle joining block of polystyrene
were complexed with lithium perchlorate in tetrahydrofuran and
cast into films (167). The polystyrene block markedly improved
film-forming characteristics and mechanical properties, but low-
ered ion conductivity. Addition of methyl tetraethylene glycol
increased ion conductivity by a factor of about five. The synthesis and preparation of other polymers useful for complexing
cations have been described (168).

Poly(1-proline)/poly(ethylene oxide)/poly(1-proline) water-
soluble ABA block copolymers have been prepared from amine-
terminated poly(ethylene oxide) and 1-proline-N-carboxyanhy-
dride (169). In this polymer, the poly(amino acid) is in a
helical conformation and the poly(ethylene oxide) is in a random
coil. The polymer forms what appears to be 1:1 complexes with
poly(methacrylic acid).

B. Poly(propylene oxide) (170)

Moacanin and Cuddihy (171) were prompted to study high- and
low-molecular-weight poly(propylene oxide) in combination with
lithium perchlorate because they noticed that marked negative
deviations from additivity were exhibited by the combinations'
specific volume. This indicated the existence of strong inter-
active forces. The high-molecular-weight polymer used was of
the propylene oxide/butadiene monoepoxide copolymer type (see
Ref. 97, Chapter 5). Dynamic mechanical loss, volumetric, and
modulus determinations as a function of temperature revealed
that the glass transition temperature, T_g, of the combinations
was markedly increased over that of the poly(propylene oxide)
alone. In the case of a 2000-molecular-weight polymer containing 25 percent lithium perchlorate, T_g was increased by 110°C.
The changes were attributed to strong dipole interactions of
Li^+ with the propylene oxide units, which reduced the segmental mobility of the polymer chains.

Wetton et al. first investigated the complexation of low- and
high-molecular-weight polymer with cobalt-II chloride and zinc
chloride (172). The high-molecular-weight polymer used in
these studies contained copolymerized allyl glycidyl ether units
(see Ref. 95, Chapter 5). Although complexes formed with
both salts, cobalt-II chloride increased only the T_g of low-
molecular-weight polypropylene glycol, whereas zinc chloride did
not discriminate by molecular weight. In these systems, the
cobalt complex involved both intermolecular and intramolecular
coordination, and the zinc complex involved only intramolecular
coordination. These studies were later enlarged (173) to include a variety of salts—-$ZnBr_2$, ZnI_2, $FeCl_3$, $SnCl_2$, $HgCl_2$,
$CuCl_2$, and LiCl—all of which formed complexes with polyoxypropylenes. The $CuCl_2$ and LiCl complexes were different from
the others in that they were opaque, were water sensitive, and
did not significantly alter T_g. All other complexes were amorphous and transparent, though colored, with well-defined T_gs,
with an increase in the case of the cobalt system of 140°C over
that of the original polymer. It was postulated that in the complex, the cations were interacting with two chain ether oxygens
to form a five-membered ring. Studies with poly(tetramethylene
oxide) were also carried out.

Stevens and Schanz (174) used differential scanning calorimetry to study the increase in T_g for complexes of a 4000-
molecular-weight polypropylene glycol and lithium perchlorate,
lithium triflate, and sodium triflate at various salt concentra-

tions. The T_g increased linearly from about $-75°C$ to about
$10°C$ with increasing salt concentration up to a 7:1 ratio of
ether oxygen to alkali metal atom. The increase was independent of the alkali metal salt used. Except for lithium perchlorate,
which remained linear to a 6:1 ratio, below the 7:1 ratio there
appeared to be an approach to an asymptotic value for T_g,
with the lithium perchlorate complex having the highest temper-
ature-limiting value. The cations are essentially free, because
of poor screening by the anions, to interact strongly with the
polarizable ether-oxygen atoms in the backbone and thus cause
volume contractions. This causes stiffening and the linear increase in T_g. Above the characteristic ether:metal ratios, the
anions become effective and decrease the ability of the cations
to further markedly increase their forces on the ether-oxygen
atoms.

Lithium and sodium salts have been complexed with propylene oxide/ethylene oxide block copolymers. Conductivity was
markedly increased in the complexes over that of the polymers,
with the greatest increases occurring at low salt concentrations
where the salt is mainly increasing T_g (175). Another study
conducted in nonaqueous solution indicated that conductivity in
the block copolymer complex, as well as in other complexes,
was affected by the size of the metal cation and the nature of
the solvent in which the complex was formed, as well as by
polymer composition and structure (176). A block copolymer
prepared by coupling ethylenediamine and poly(ethylene glycol)
with 4,4'-diphenylmethane diisocyanate and doped with lithium
perchlorate yielded high ionic conductivity (177).

An interest in polyurethanes that have good blood compatibility and are used in cardiovascular and vascular repair devices led to an investigation of the interaction of propylene oxide polyol-based polyurethane-ureas with lithium bromide (178).
Although such polyurethane-ureas are considered to be hydrophobic in nature and sorb only about 2 percent water, addition
of small amounts of lithium bromide resulted in highly significant increases in water sorption. Water sorption increased by
more than 10 times the weight of the added salt up to a salt
level (6 to 18 percent) that depended on the polyol molecular
weight or hard-segment content of the polyurethane-urea.
Above this level, the amount of water sorbed was much lower
and was essentially one-to-one with added salt. These effects
were demonstrated to be linked to the salt; specifically, linking
to urea groups in the hard segment that destroyed interchain

hydrogen bonding and caused the polymer to lose its typical domain structure and become amorphous.

Polyurethane networks were prepared by crosslinking ethylene oxide/propylene oxide block copolymers, ethylene oxide/propylene oxide/dimethyl siloxane graft copolymers, and poly-(ethylene oxide) with various isocyanates (179). Complexation with lithium perchlorate led to network shrinkage and increases in T_g. The data suggested regular chain partitioning by the salt.

C. Polyepichlorohydrin

Wissbrum and Hannon have used variations in T_g to investigate polymer-salt interactions in a number of polymers (180). One of their studies involved complexation of a copolymer prepared from 2,2-bis-4-hydroxyphenylpropane and epichlorohydrin with calcium thiocyanate (181). The salt caused an increase in T_g that was related to reduction in the free volume of the polymer.

IV. INTERFACIAL CHARACTERISTICS

Many applications for poly(ethylene oxide) involve its use at or on surfaces, and its interfacial characteristics are important. This polymer has been shown to be very surface active in aqueous solutions (182—186). When adsorbed onto an interface from dilute solution, the molecular coil of poly(ethylene oxide) takes on a two-dimensional, flattened conformation with a majority of the molecular segments in contact with the interface (187, 188). As the molecular weight increases, the fraction of the molecule oriented at the interface decreases. This change is related to the relative domination of enthalpic and entropic factors.

When liquid-liquid, hydrophobic-hydrophilic interfaces exist, the poly(ethylene oxide) molecule can have different conformations in the different phases. In the case of solubility in both phases, the helical or more ordered conformation exists in the water-phase boundary and a more random conformation exists in the organic-phase boundary.

Sauer and Yu (189) examined the adsorption kinetics of poly(ethylene oxide) at the air-water interface. The viscoelastic parameters of the films formed by spontaneous adsorption from aqueous solution are same as those formed by spreading from a methylene chloride solution. The early and first adsorption kinetic data of Glass (183) were reanalyzed, and it was found that the diffusion values were actually larger only by a factor of about two from the bulk solution values, rather than the several orders of magnitude increase reported earlier. The difference was attributed to convective problems existing in the early work.

Studies of the adsorption onto various substrates such as mica (190–192), silica (193, 195), aqueous magnesium chloride and sulfate solutions (196), porous media (197), and polystyrene spheres (198) and latexes (199) has been done.

V. BLENDS WITH OTHER POLYMERS

Polymer blends are interesting materials from both fundamental and product-development standpoints. Such systems offer a creative way to specifically design products for particular applications. If polymer blends are viewed in a simplistic way, the free energy of mixing, ΔG, which is given by

$$\Delta G = \Delta H - T\Delta S$$

must be negative for miscibility to take place. Since molar quantities are involved and since, by their very nature, polymers are high-molecular-weight entitites, small values are being dealt with. The positive entropy change that occurs when polymers are mixed favors a negative free-energy change and miscibility; but the change is very small, and unfavorable heats of mixing, ΔH, are usually positive and greater; thus, immiscibility is the expected result when high polymers are blended. Miscibility usually requires entropic stabilization. The possibility of miscible systems occurring is enhanced as molecular weight decreases and temperature is increased.

Recent theories of polymer blending are concerned with three main contributions to the free energy of mixing or blending (200–203). These contributions of combinational entropy, exchange interactions, and free-volume contributions are delicately balanced, and miscibility or immiscibility can be altered by small changes in these factors. As might be expected, such behavior can lead to relatively complex phase diagrams that are similar to those encountered in metallurgy and alloy formation.

Polyoxyethylene has been blended with a number of polymers. In the case of poly(methyl methacrylate), PMMA, a num-

ber of studies have demonstrated miscibility of the polymers in
the amorphous state when the PMMA is atactic (204–215). Silvestre and coworkers (213) investigated the effect of PMMA tacticity on the balance of entropy, interaction, and free volume of
mixing or miscibility and superstructure of the blends. The
long period and the amorphous and transition region thicknesses
were independent of composition when isotactic PMMA was used.
However, for atactic and syndiotactic PMMA, these characteristics increased with increasing PMMA content. Glass transition
and melting data indicated that the systems containing isotactic
PMMA are phase separated in the amorphous state, whereas
those based on syndiotactic or atactic PMMA are homogeneous,
one-phase systems. The differences in behavior are related to
the free-volume contributions to the free energy of mixing.

Kim and Porter (216) investigated the uniaxial drawing of
high-molecular-weight poly(ethylene oxide)/PMMA blends. As
the poly(ethylene oxide) content was decreased, both blend
crystallinity and uniaxial ductility were markedly decreased from
a draw ratio of 36 for poly(ethylene oxide) alone (217) to 5 for
a 40/60 blend of polyether and PMMA. At high polyether content, the poly(ethylene oxide) domains are of a semicrystalline
nature. In addition, the melting temperature was found to be
difficult to interpret in terms of interaction parameters. Significant entropic effects were thought to occur during the drawing
process, which involves a solid-state coextrustion from a capillary rheometer. Zhao and coworkers (218) studied similarly
prepared blends using infrared dichroism and birefringence as
analytical tools. Mechanical properties and master relaxation
curves at $(T_g + 50)°C$ were developed for compatible blends
containing up to 20 percent poly(ethylene oxide).

Poly(ethylene oxide)/poly(vinyl acetate) blends (219, 220)
are miscible in the melt over the entire composition range. The
interaction parameter is negative, which indicates thermodynamic
stability in the melt. Melt-quenched, solid-state blends are
miscible at up to 40 percent weight fraction poly(vinyl acetate),
PVAC. In blends containing up to 25 percent PVAC, the crystallization of poly(ethylene oxide) is essentially unhindered over
that over the polymer alone. The melting-point depression of
the polyether is related to both morphological and thermodynamic factors. Glass transition data supported the findings.
The interfacial activity of poly(ethylene oxide) and PVAC at
air/water and at oil (heptane)/water interfaces has been studied
by surface light scattering (221). Similar studies were carried

out for an ethylene oxide/styrene block copolymer (222). The results of these studies should be useful in the investigation of blends.

Blends of poly(ethylene oxide) and poly(vinyl chloride) containing up to 50 percent of the latter polymer have a negative interaction parameter, which indicates that the blends have melt compatibility (223). As the poly(vinyl chloride) content increases, the equilibrium melting point decreases in the mixtures. Analysis of the change in melting point indicated that entropic effects made little contribution to the interaction parameter.

When low-molecular-weight polyoxyethylene was blended with polysulfone, homogeneous blends at temperatures between the melting point and the cloud point of the polyether could be obtained only when the blend contained 40 percent or more polysulfone (224). Increasing the polyether molecular weight resulted in inhomogeneity over the entire compositional range. When blended with polyethersulfone, miscible blends were obtained. Interest in such blends stems from the fact that polyoxyethylene is used as a pore-forming additive or as a means of increasing viscosity when the engineering polymers are produced as hollow fiber membranes and asymmetric plane sheets.

Polyoxyethylene has been examined as a polymer that could be used to improve the processibility and toughness properties of poly-d(-)-3-hydroxybutyrate (225). The latter optically active polymer, which is prepared by biosynthesis (226), has a high melting point of 180°C, is truly biodegradable and very biocompatible, and has excellent gas-barrier properties. It has promise as a packaging material, except that the homopolymer is highly crystalline and brittle in character. Blends with polyoxyethylene have negative interaction parameters and, as would be expected, are miscible in the melt. Crystallized melts exhibit a single glass transition temperature, and there is a depression of the equilibrium melting temperature. Below the apparent melting points of the polymers, it appears that interfibrillar and/or interlamellar amorphous and homogeneous mixtures of the two polymers exist.

Blends of polyethylene glycols having molecular weights of 1500 to 35,000 and poly(lactic acid) were studied with infrared spectroscopy and thermal analysis (227). When the blends contained more than 20 percent of either component, they could crystallize and consisted of two semimiscible crystalline phases that were dispersed in an amorphous phase. In blends containing less than 20 percent of either component, only the com-

ponent present in the larger amount was able to crystallize, and
the amorphous matrix consisted of the minor component and the
amorphous phase of the major component. Molecular weight of
the polyether had an effect on morphology, with larger degrees
of crystallinity arising as molecular weight increased.

A variety of other blends have been examined, including
mixtures of various molecular weight polyethylene glycols (228),
styrene/methacrylic acid ionomer in combination with polyoxyethylene or polyoxypropylene (229), methyl methacrylate/
methacrylonitrile copolymers and methyl methacrylate/glycidyl
methacrylate copolymers blended with polyepichlorohydrin (230).
Poly(ethyl methacrylate) and poly(ethylene oxide) (231), as well
as poly(dimethyl siloxane) with poly(ethylene oxide) (232), also
form miscible blends. Polyepichlorhydrin and poly(ethylene oxide) form completely compatible binary blends, and polyeipi-
chlorohydrin, poly(ethylene oxide), and poly(methyl methacrylate) form completely miscible ternary blends (233). Poly-
(bisphenol A hydroxy ether) also forms such miscible blends
with poly(ethylene oxide) (234). Useful information regarding
the interfacial tension of mixtures of polyethylene glycol and
polypropylene glycol have been published (235).

Although not polymer blends, two-component, linked networks have been prepared from functionalized polyoxyethylene
and polybutadiene (236, 237). By linking the two polymers,
a degree of compatibility is forced, since the molecules are no
longer entirely free of phase separation. The systems demonstrated that the hydrophilic-hydrophobic balance could be altered by varying the polyether chain lengths and the crosslink
density.

VI. DIELECTRIC BEHAVIOR

The molecular dynamics, such as flexibility and inter-intramo-
lecular interactions, and the electrical properties of polar molecules like the poly(alkylene oxide)s can be investigated by
measurement of simple and complex dielectric phenomena. From
dipole relaxation times, the time and temperature dependence of
polymer flexibility and mobility or viscoelasticity in bulk and in
solution, which is important to flow characteristics and utility,
can be analyzed and studied. It is well known that complex
mechanical moduli are analogous to dielectric phenomena.

Marchal and Benoit (238) studied the dipole moments of
polyethylene glycols from dimer to a molecular weight of 10,000.

The mean-square dipole moment decreased rapidly as the molecular weight increased and approached a limiting value of 1.13. Thus, small segments of the polymer chain orient in an applied electrical field. Theoretical and experimental values of the dipole moments were in fair agreement when short-chain interactions and steric hindrances were taken into account. The dipole moment was independent of excluded volume effects, indicating that the short-range interactions probably can be evaluated in nontheta solvents.

Dipole relaxation times were also measured by Davies et al. (239) over a molecular weight range of about 200 to 29,000 in bulk and in aqueous and benzene solutions. The distribution factors for the dipole relaxation times in benzene at 20°C were independent of molecular weight up to a value of 4000, indicating that this parameter is an attribute of the molecular chains and not their size or environment. There was little change in the relaxation process when going from dilute solution to the bulk state, reflecting the marked reorientation freedom in these polymers. The evidence suggested that the carbon-oxygen linkage acts as a ball-and-socket joint, making the average freely jointed unit little more than a single monomeric unit. Additional confirmation that small segments of long molecular chains reorient in the electrical field was given by the tendency toward constant values for the dielectric absorption and dielectric constant as the molecular weight increased. Information from the aqueous-solution studies indicated that each monomer unit was acting independently and interacted with water to about the same extent as a dioxane molecule.

From data obtained in the microwave frequency region, the temperature dependence and the activation energies for polyoxyethylene in benzene were obtained. At a molecular weight of 4000, the enthalpy change was 11.30 ± 0.42 kJ/mol (2.7 ± 0.1 kcal/mol), and the entropy change was 2.51 ± 0.84 kJ/mol K (0.6 ± 0.2 kcal/mol K). At a molecular weight of 29,000, the enthalpy change was 10.04 ± 0.42 kJ/mol (2.4 ± 0.1 kcal/mol), and the entropy change was essentially 0. It was concluded that the large differences observed between the concentration dependence of the viscosity and the dielectric relaxation times were a reflection of the different configurational or entropy changes involved in the two processes.

The dielectric behavior of polyethylene glycols at 25°C was also investigated by Magnasco and coworkers (240), who obtained the orientation polarizations at zero concentration and the dipole moments. Although their results were not in agreement with the hypothesis of free rotation, the experimental values of specific polar moments were in agreement with statistically calculated values that assumed hindrance of bond dipole rotation by a potential barrier that is asymmetric around the trans configuration. Other investigators (241) studied dipole properties over a broad range of molecular weights and obtained results that were in agreement with the preferred gauche configuration for the $-CH_2-CH_2-$ group rotational state and the trans configuration for the CH_2-O and $-CH_2-O$ groups.

Porter and Boyd (242) conducted detailed investigations of the alpha, beta, and gamma relaxations and found that only the latter two were active. The beta relaxation is associated with amorphous regions, since it is evident in both solid and molten polyoxyethylene. The beta relaxation was also investigated by Connor et al. (243) in the microwave region for bulk polymers over the molecular weight range of 4000 to 2.8×10^6. The gamma relaxation has been observed in single crystals (244). The dielectric properties of polymers at microwave frequencies, including brief mention of the poly(alkylene oxide)s, have been reviewed recently (245).

Baur and Stockmayer (246) and Williams (247) investigated the dielectric properties of bulk, atactic polypropylene glycols as a function of temperature and molecular weight. At a given temperature, the main dispersion occurred at virtually the same frequency for all molecular weights. At low frequencies, a second, small dispersion maximum that was strongly dependent on molecular weight was apparent. It was attributed to a cumulative dipole contribution whose magnitude was dependent on long-range backbone motions or conformations. Two relaxations are expected, since there are two dipole components in operation in poly(propylene oxide). One component is parallel and the other is perpendicular to the chain contour (248). These relaxation processes were examined by Mashimo et al. (249) in dilute benzene solution over the frequency range of 10 MHz to 10 GHz. The results were similar to those found earlier. The low-frequency process was strongly dependent on molecular weight and was well explained by the conformational theories of Rouse (250) and Zimm (251). The second or high-frequency process, which was molecular weight independent, was caused by an elementary relaxation process in backbone motion such as rotational motion of a chain bond. The longer relaxation time, smaller dipole moment ratio, and larger activation energy of polyoxypropylene when compared to polyoxyethylene probably indicate more resistance to relative bond rotation due to the methyl group.

The addition of water to a 4000-molecular-weight polypropylene glycol caused an antiplasticization of the polymer that was manifested by a shift of the beta-relaxation maximum to a higher temperature (252). The effect was attributed to increased effective chain length through linkage of chain end groups by hydrogen bonding with water molecules. In a second study (253), dielectric loss and permittivity were measured for propylene glycol and a 425-molecular-weight polypropylene glycol over a frequency range of 5 to 12×10^5 Hz and a temperature range of 77 K to 320 K. Two relaxation maxima were apparent in the polymeric material, but only one was apparent in the glycol. It appears that the second and smaller maximum is masked by the large contribution to the loss from the main relaxation. Water (up to 1.25 percent) added to the 425-molecular-weight polymer increased the temperature of the main relaxation and decreased the magnitude of the secondary process, which is related to less efficient packing that occurs as molecular weight is increased. In propylene glycol, water increases the half-width of the main relaxation.

Blythe and Jeffs (254) investigated the dielectric-relaxation behavior of essentially amorphous, high-molecular-weight polyepichlorohydrin and polyepibromohydrin prepared with a zirconium/acetylacetonate/triethyl aluminum catalyst. The investigation was carried out over a wide temperature range and a frequency range of 10^{-1} to 10^4 Hz. Two distinct relaxation processes were apparent in the dielectric-relaxation spectra, with the high-temperature relaxation related to the glass-to-rubber transition and the low-temperature relaxation related to rotational motion of the chloromethyl and bromomethyl side groups. Two particular characteristics were noticed in the spectra. The relaxation maxima were very broad, and the activation energy was greater for the smaller chloromethyl group than for the larger bromomethyl group. These effects were explained by ascertaining that the intramolecular dipole-dipole interactions are weak and that the intermolecular dipole-dipole interactions account for the observed activation-energy orders.

VII. GLASS TRANSITION TEMPERATURE AND MECHANICAL PROPERTIES

The temperature at which an amorphous polymer or the amorphous portion of a partially crystalline polymer is transformed

TABLE 5   Glass Transition Temperatures for Various Poly(alkylene oxide)s

Polymer	Tg, °C	Method	Ref.
Poly(1,2-butene oxide)	− 69	DTA	255
Polyepibromohydrin	− 14	DTA	254
Polyepichlorohydrin	− 22	DTA	254
Poly(ethylene oxide)	− 52	Mechanical loss	256
Poly(propylene oxide)	− 67	DTA	255

from a glassy state to a rubbery state is termed the glass transition temperature and is denoted by T_g. At this temperature,
the modulus of a polymer undergoes marked changes of about 3
to 5 orders of magnitude, and the mechanical loss, viscous component of the complex shear modulus, dielectric loss, etc. have
maxima. T_g can be determined by a variety of techniques,
including mechanical, dielectric, volumetric, and calorimetric.
T_g is dependent on frequency of measurement (i.e., it is a sec-
ond-order relaxation phenomenon), molecular weight, rate of
heating, and other factors such as crystallinity. Because of
such dependencies, various values of T_g can be found in the
literature for the same polymer. Table 5 is a tabulation of T_g
values for the poly(alkylene oxide)s discussed in this book. Lal
and Trick (255) have also presented values for other 1,2-olefin
epoxides and sulfides.

Usually, T_g increases as a function of number average molecular weight, M_n, in accord with the relationship

$$(1/T_g) = (1/T_g) + (K/M_n)$$

where T_g is the glass transition temperature of an "infinite"-
molecular-weight polymer and K is a constant. It has been
shown in the literature (256, 257) that poly(ethylene oxide)s do
not behave in accordance with this relation, and much higher
values than expected are obtained experimentally for polymers
with molecular weights of less than 1000 to about 100,000. A
maximum value of about −20°C is obtained at a molecular weight
of about 10,000. It was later demonstrated that this apparent

TABLE 6   Tensile Properties of Poly(ethylene oxide)   (From Ref. 267.)

Ultimate tensile strength, psi	1800−2400	(12.4−16.5 MPa)
Yield strength, psi	1000−1500	(6.9−10.3 MPa)
Percent ultimate elongation	700−1200	
Percent elongation at yield	5−30	
Durometer hardness, Shore A	99	

anomaly was due to crystallinity; when well-vitrified, amorphous specimens were studied, the expected relationship was obtained (258).   Other investigators (259) found that T_g was independent of molecular weight over the range of 3000 to 22,000 and had a value of −60°C, and still others found that thermal prehistory had a definite effect on T_g (260, 261).   The relaxation behavior of polyoxyethylene has been reviewed (262).

Polyoxypropylenes have been shown to have glass transition temperatures that are relatively independent of molecular weight over almost the entire molecular weight range when various methods of measurement are used (263−266).   Recently, Johari and coworkers (267) investigated the calorimetric relaxation of polypropylene glycols as well as propylene glycol and found that the increase in T_g with molecular weight is much more rapid than has been found with other polymers.   A limiting T_g value—that of an "infinite"-molecular-weight polymer—was obtained for a molecular chain containing 20 monomer units.   The relaxation behavior of polyoxypropylene has been reviewed (262).

The tensile properties of poly(propylene oxide) and polyepichlorohydrin elastomers were presented in Tables 5.1 and 5.2, respectively.   The tensile properties of high-molecular-weight poly(ethylene oxide) are given in Table 6.   These properties can be altered markedly when the relative humidity is greater than 70 to 80 percent.   For example, at 90 percent relative humidity, the tensile modulus decreases from 50,000 psi (345 MPa) to 1000−5000 psi (6.9−34.5 MPa) (268).   Ultradrawn filaments of different molecular weight poly(ethylene oxide)s have been investigated as a function of uniaxial draw ratio (269, 270).

REFERENCES

1. F. E. Bailey, Jr., and J. V. Koleske, Poly(ethylene oxide), Academic Press, New York, 1976.
2. F. E. Bailey, Jr., and J. V. Koleske, in Nonionic Surfactants, Physical Chemistry (M. J. Schick, ed.), Marcel Dekker, New York, 1987, p. 927.
3. V. I. Klenin, N. K. Kolnibolotchuk, and N. A. Solonina, Vysokomol. Soedin. Ser. A, 30 (10): 2076 (1988); C.A. 110: 58539 (1989).
4. K. L. Smith and A. E. Winslow, unpublished results, 1956.
5. G. M. Powell III and F. E. Bailey, Jr., in Kirk-Othmer, Encyclopedia of Chemical Technology, Second Supplement Volume, Wiley-Interscience, New York, 1960, p. 597.
6. P. M. Galin, Polymer 24:323 (1983).
7. F. E. Bailey, Jr., G. M. Powell III, and K. L. Smith, Ind. Eng. Chem. 50:8 (1958).
8. F. E. Bailey, Jr., and R. W. Callard, J. Appl. Polymer. Sci. 1:56 (1963).
9. M. Rosch, Koll. Z. 147:78 (1956); Fette Seifern Anstrichmittl. 65:223 (1963).
10. Y. Kambe and C. Honda, Polymer Communications 25:154 (1984).
11. J. Roots and B. Nystrom, Chimica Scripta 15:165 (1980).
12. W. F. Polik and W. Burchard, Macromolecules 16:978 (1983).
13. G. Bazzi, P. L. Johnson, and E. Gulari, "Conformational Properties of Aqueous Poly(ethylene oxide) Solutions," 58th Colloid and Surface Science Symposium, June 10–13, 1984, Carnegie-Mellon University, Pittsburgh.
14. H. S. Harned and B. B. Owen, Physical Chemistry of Electrolytic Solutions, 2nd ed., Reinhold, New York, 1950, p. 397.
15. F. Hofmeister, Arch. Exp. Natur. Pharmakrol. 24:247 (1888); 27:395 (1890).
16. M. J. Schick, J. Coll. Sci. 23:801 (1962).
17. T. C. Amu, Polymer 23:1775 (1982).
18. W. H. Stockmayer and M. Fixman, J. Polymer Sci. C1: 137 (1963).
19. P. J. Flory, Principals of Polymer Chemistry, Cornell University Press, Ithaca, 1953.
20. J. M. Harris, N. H. Hundley, T. G. Shannon, and E. C. Struck in Crown Ethers and Phase Transfer Catalysis in

Polymer Science (L. Mathias and C. E. Carreher, eds.),
Plenum Press, New York, 1984, p. 371.

21. K. P. Ananthapadmanabahn and E. D. Goddard, Langmuir
    3:25 (1987).
22. P. J. Flory, Proc. Roy. Soc. (London) A234:73 (1956).
23. J. W. McBain, Colloid Science, Heath, Boston, 1950.
24. C. K. Chiang, G. T. Davis, C. A. Harding, and J.
    Aarons, Solid State Ionics, :1121 (1983).
25. G. T. Davis and C. K. Chiang, in Conducting Polymers
    (H. Sasabe, ed.), Chemical Marketing Center, Tokyo,
    1984, pp. 1—22.
26. K. Durham, Surface Activity and Detergency, MacMillan,
    London, 1961.
27. J. E. Weston and B. C. Steele, Solid State Ionics 2:347
    (1981).
27a. R. L. Peck, U.S. Patent No. 4,797,190 (1989).
28. C. Robitaille, S. Marques, D. Boils, and J. Prud'homme,
    Macromolecules 20:3023 (1987).
29. E. A. Boucher and P. M. Hines, J. Polymer Sci.:
    Polymer Phys. Ed. 16:501 (1978).
30. D. Boils-Boissier, ACS Proceedings (Spring Meeting,
    Dallas), Polymers Material Science and Engineering Div.,
    Vol. 60 (1989).
31. T. K. Dzhumadilov, S. Ismagulova, and E. A. Bekturov,
    Izv. Akad. Nauk. Kaz. SSR, Ser. Khim. 6:22 (1988);
    C.A. 110: 115459 (1989).
32. K. Ono, H. Honda, and K. Murakami, J. Macromol. Sci.-
    Chem. A26:567 (1989).
33. D. Teeters, S. L. Steward, and L. Svoboda, Solid State
    Ionics 28—30 (Pt. 2):1054 (1987).
34. A. H. Kovacs, B. Lotz, and A. Keller, J. Macromol.
    Sci.-Phys. B3 (3):385 (1969).
35. P. H. Geil, Polymer Single Crystals, Wiley Interscience,
    New York, 1963.
36. A. Keller, in Growth and Perfection of Crystals (R. H.
    Doremus, et al., eds.), John Wiley & Sons, New York,
    1958, p. 1.
37. W. J. Barnes and F. P. Price, Polymer 5:283 (1964).
38. F. N. Hill, F. E. Bailey, Jr., and J. T. Fritzpatrick,
    Ind. Eng. Chem. 50:5 (1958).
39. E. R. Walter and F. P. Reding, Proceedings of American
    Chemical Society, San Francisco (1958).
40. H. F. White and C. M. Lovell, J. Polymer Sci. 41:367
    (1959).

41. H. Matsura and T. Miyazawa, Bull. Chem. Soc. Japan
    41:1798 (1968).
42. H. Matsurra and T. Miyazawa, J. Polymer Sci. Part A2
    7:1735 (1969).
43. E. F. Oleiniki and N. S. Enikolopyan, J. Polymer Sci.
    Part C 15:3677 (1968).
44. V. P. Roshchupkih, B. V. Ozerkovskii, and G. V.
    Korolov, Vysokomol. Soyed., A11 (6):1366 (1966).
45. J. L. Koenig and A. C. Angood, J. Polymer Sci. Part A2
    8:1787 (1970).
46. J. F. Rabot, K. W. Johnson, and R. N. Zitter, J. Chem.
    Phys. 61 (2):504 (1974).
47. T. Yoshihara, H. Tadakoro, and S. Murabashi, J. Chem.
    Phys. 61 (2):504 (1974).
48. H. Tadakoro, Y. Takahashi, and Y. Chatani, Makromol.
    Chem. 109:96 (1967).
49. T. Miyazawa, K. Fumushima, and Y. Iideguchi, J. Chem.
    Phys. 37:2764 (1962).
50. T. Miyazawa, J. Chem. Phys. 35:693 (1961).
51. W. J. Barnes, W. G. Luetzel, and F. P. Price, J. Phys.
    Chem. 65:80 (1961).
52. F. P. Price, J. Polymer Sci. 3A:3079 (1965).
53. J. Schultz, Polymer Material Science, Prentice-Hall,
    Englewood Cliffs, New Jersey, 1974.
54. M. H. Thiel, J. Polymer Sci.: Polymer Phys. Ed. 13:1097
    (1975).
55. D. C. Bassett, Principles of Polymer Morphology, Cam-
    bridge University Press, Cambridge, 1981.
56. N. Ding, R. Salovey, and E. J. Amis, Polym. Prepr.
    (Am. Chem. Soc., Div. Polym. Chem.) 29 (2):385 (1988).
57. A. J. McHugh, W. R. Burghardt, and D. A. Holland,
    Polymer 27:1585 (1986).
58. M. J. Avrami, J. Chem. Phys. 1:1103 (1939); 8:212
    (1940); 9:117 (1941).
59. J. D. Hoffman and J. I. Lauritzen, J. Res. Natl. Bur.
    Stand. 65A:297 (1961).
60. J. D. Hoffman, SPE Trans. 4 (4):78 (1964).
61. F. Gornich and J. D. Hoffman, Ind. Eng. Chem. 58 (2):
    41 (1966).
62. Yu. K. Godovsky, G. L. Slonimsky, and N. M. Garbar,
    J. Polymer Sci., Part C 38:1 (1972).
63. C. J. Murphy and G. V. S. Henderson, Jr., Polymer
    Eng. Sci. 27 (11):781 (1987).

64. J. D. Hoffman and J. J. Weeks, J. Res. Nat. Bur.
     Stand. 66A:13 (1962).
65. L. Mandelkern, J. Polym. Sci. 47:494 (1960).
66. J. D. Hoffman, Macromolecules 19:1124 (1986).
67. W. Banks and A. Sharples, Macromol. Chem. 59:233
     (1963).
68. I. H. J. Hillier, J. Polymer Sci. Part A2 4:1 (1966).
69. L. Mandelkern, Crystallization of Polymers, McGraw Hill,
     New York, 1984.
70. A. Sharples, Introduction to Polymer Crystallization,
     Arnold, London, 1963.
71. D. R. Beech, C. Booth, D. V. Dodgson, and I. H.
     Hillier, J. Polymer Sci. Part A2 10:1555 (1972).
72. J. Q. G. Maclaine and C. Booth, Polymer 16:680 (1975).
73. S. Z. D. Cheng and B. Wunderlich, J. Polymer Sci.
     Part B: Polymer Physics 24:577 (1986).
74. S. Z. D. Cheng, H. S. Bu, and B. Wunderlich, J.
     Polymer Sci. Part B: Polymer Physics 26:1947 (1988).
75. K. Song and S. Krimm, Macromolecules 22 (3):1504
     (1989).
76. S. Z. D. Cheng, D. W. Noid, and B. Wunderlich, J.
     Polymer Sci. Part B: Polymer Phys. 27:1149 (1989).
77. N. Ding, E. J. Amis, M. Yang, and R. Salovey, Polymer
     29:2121 (1988).
78. M. J. Fernandex-Berridi, G. M. Guzman, and J. J.
     Iruin, Makromol. Chem. 184 (3):605 (1983).
79. M. Galin, Eur. Polymer J. 19:1 (1983).
80. E. Calahorra, M. Cortazar, and G. M. Guzman, Polymer
     Comm. 24:211 (1983).
81. E. Martuscelli, C. Silvestre, M. L. Addonizio, and L.
     Amelino, Makromol. Chemie 187:1557 (1987).
82. D. M. Hoffman, Thesis, University of Massachusetts,
     Amherst (1979).
83. L. Xuan, W. Yinkang, and Y. Yang, Chinese J. Polymer
     Sci. 3:108 (1988).
84. C. Wohlfarth, E. Regener, and M. T. Ratzsch, Makromol.
     Chem. 190:145 (1989).
85. A. V. Filipov, V. S. Smirnov, R. S. Gimatdinov, and
     Y. D. Shibanov, Vysokomol. Soedin. Ser. B 30 (11):854
     (1988); C.A. 110: 58448 (1989).
86. X. Li, Shiyou Huagong 17 (6):382 (1988); C.A. 109:
     231853 (1988).
87. A. Cholli, F. C. Schilling, and A. E. Tonelli, Polym.

Prepr., Am. Chem. Soc., Div. Polymer Chem. 29 (1):
12 (1988).

88. F. E. Bailey, Jr., and J. V. Koleske, J. Chem. Ed. 55:
     761 (1973).
89. E. Stanley and M. Litt, J. Polymer Sci. 43:453 (1960),
90. D. G. Natta, P. Corradini, and G. Dall'Asta, Atti.
     accad. nazl. Lincei Rend. classe sci. fis. mat. nat. 20:
     408 (1956).
91. C. C. Price, M. Osgan, R. E. Hughes, and C. Shamblin,
     J. Am. Chem. Soc. 78:690 (1956).
92. R. L. Miller and L. E. Nielsen, J. Polymer Sci. 44:391
     (1960).
93. L. E. St. Pierre and C. C. Price, J. Am. Chem Soc. 78:
     3432 (1956).
94. G. Allen, Soc. Chem. Ind. Monograph 17:167 (1963).
95. R. N. Work, R. D. McCammon, and R. G. Saba, Bull.
     Am. Phys. Soc. 8:266 (1963).
96. G. Allen, C. Booth, and M. N. Jones, Polymer 5:195
     (1964).
97. M. Cesari and G. Perego, Makromol. Chem. 133:133
     (1970).
98. R. E. Hughes and R. J. Cella, Polymer Preprints, Am.
     Chem. Soc. Div. Polymer Chem. 15:137 (1979).
99. B. Trzebicka and E. Turska, Polymer 29:1689 (1988).
100. H. Janeczek, B. Trzebicka, and E. Turska, Eur. Polymer
     J. Comm. 28:123 (1987).
101. B. Trzebicka and E. Turska, Polymer 26:387 (1985).
102. B. Trzebicka, S. Smigaziewicz, and E. Turska, Eur.
     Polymer J. 27:1067 (1986).
103. E. Tsuchida and K. Abe, Interactions Between Macromole-
     cules in Solution and Intermacromolecular Complexes, Ad-
     vances in Polymer Sci. Ser. No. 45, Springer-Verlag,
     New York, 1982, pp. 1–125.
104. K. L. Smith, A. E. Winslow, and D. E. Peterson, Ind.
     Eng. Chem. 51:1361 (1959).
105. G. E. Barker and J. H. Ronsanto, J. Am. Oil Chem.
     Soc. 32:249 (1955).
106. F. E. Bailey, Jr., and H. G. France, J. Polymer Sci.
     49:397 (1961).
107. H. Mima, Yukagaku Zasshi 79:857 (1959); C.A. 53:
     20846 (1959).
108. F. E. Bailey, Jr., R. D. Lundberg, and R. W. Callard,
     J. Polymer Sci. Part A 2:845 (1964); J. Polymer Sci.
     Part A 4:1563 (1966).

109. S. Saito, in Nonionic Surfactants: Physical Chemistry (M. J. Schick, ed.), Marcel Dekker, New York, 1987, p. 881.
110. H. Tadakoro, T. Yoshihara, Y. Chatani, and S. Murahashi, J. Polymer Sci. Part B 2:363 (1964).
111. P. V. Wright, J. Macromol. Sci.-Chem. A26 (2&3):519 (1989).
112. P. V. Wright, Polymer 30 (6):1179 (1989).
113. A. A. Blumberg, S. S. Pollack, and C. A. J. Hoeve, J. Polymer Sci. Part A 2:2499 (1964).
114. H. Tadokoro, R. Iwamoto, and Y. Saito, Annual Mtg. Soc. of Polymer Sci. (Japan), Nagoya (May 1961).
115. J. Parrod and A. Kohler, Compte Rend. 246:1046 (1958).
116. R. M. Myasnikova, E. F. Titova, and E. S. Obolonkova, Polymer 21:403 (1980).
117. F. Francois, J. Dayantis, and J. Sabbadin, Eur. Polymer J. 21:165 (1985).
118. H. G. Elias, Angew Chem. 73:209 (1961).
119. S. K. Chatterjee and K. R. Sethi, J. Polymer Sci.: Chem. Ed. 21:1045 (1983).
120. S. K. Chatterjee and K. R. Sethi, Polymer Comm. 24:253 (1983).
121. S. K. Chatterjee and K. R. Sethi, J. Macromol. Sci.-Chem. A-21:765 (1984).
122. E. Kokufuta, A. Yokota, and I. Nakamura, Polymer 24:1031 (1983).
123. S. K. Chatterjee, N. Chatterjee, and G. Riess, Makromol. Chem. 183:481 (1982).
124. G. Staiko, P. Antonopoulou, and E. Christou, Polymer Bull. (Berlin) 21:209 (1989).
125. H. L. Chen and H. Morawetz, Eur. Polymer J. 19:923 (1983).
126. B. Bednar, H. Morawetz, and J. A. Shafer, Macromolecules 17:1634 (1984).
127. B. Bednar, Z. Li, Y. Huang, L-C. P. Chang, and H. Morawetz, Macromolecules 18: 1829 (1985).
128. H. Oyama, W. T. Tang, and C. W. Frank, Macromolecules 20:474 (1987); Macromolecules 22:1255 (1989).
129. J. J. Heyward and K. P. Ghiggino, Macromolecules 22:1159 (1989).
130. H. T. Oyama, W. T. Tang, and C. W. Frank, Am. Chem. Soc. Symposium Series 358: 422–433 (1987).
131. S. Nishi and T. Kotaka, Macromolecules 18:1519 (1985).
132. S. Nishi and T. Kotaka, Macromolecules 19:978 (1986).

133. S. Nishi and T. Kotaka, Polym. J. (Tokyo) 21 (5):393 (1989).
134. K. G. Vassilev, D. K. Dimov, R. T. Stamenova, R. S. Boeva, and Ch. B. Tsvetanov, J. Polymer Sci. Part A: Polymer Chem. 24:3541 (1986).
135. H. Ohno, H. Takinishi, and E. Tsuchida, Makromol. Chem., Rapid Comm. 2:511 (1981).
136. A. Sjoberg and G. Karlstrom, Macromolecules 22:1325 (1989).
137. M. Niwa and N. Higashi, Macromolecules 22:1000 (1989).
138. E. A. Bekturov, L. A. Bimendina, and S. A. Saltybaeva, Makromol. Chem. 180:1813 (1979).
139. S. H. Jeon and T. Ree, J. Polymer Sci. Part A: Polymer Chem. 26:1419 (1988).
140. K. Ono and H. Hoda, J. Macromol. Sci.-Chem. A26 (2&3): 561 (1989).
141. Y. Osada, J. Polymer Sci.: Polymer Chem. Ed. 15:255 (1977).
142. Y. Osada, J. Polymer Sci.: Polymer Chem Ed. 17:3485 (1979).
143. H. Ohno, H. Matsuda, and E. Tsuchida, Makromol. Chem. 182:2267 (1981).
144. F. Quina, L. Sepulveda, R. Sartori, E. B. Abuin, C. G. Pino, and E. A. Lissi, Macromolecules 19:990 (1986).
145. S. L. Maunu, T. Korpi, and J. J. Lindberg, Polymer Bull. (Berlin) 19 (2):171 (1988).
146. K. J. Liu, Macromolecules 1:308 (1968); K. J. Liu and J. E. Anderson, Macromolecules 2:235 (1969).
147. C. J. Pederson, J. Am. Chem. Soc. 89:7017 (1967); Aldrichimica Acta 4:1 (1971).
148. G. W. Gobel and H. D. Durst, Aldrichimica Acta 9:3 (1971).
149. A. Ricard, Eur. Polymer J. 15:1 (1976).
150. E. Florin, Macromolecules 18:360 (1985).
151. J. M. Parker, P. V. Wright, and C. C. Lee, Polymer 22: 1305 (1981).
152. Y. Takizawa, H. Aiga, M. Watanabe, and I. Shinohara, J. Polymer Sci.: Polymer Chem. Ed. 21:3145 (1983).
153. C. Robitaille and J. Prud'homme, Macromolecules 16:665 (1983).
154. D. J. Bannister, G. R. Davies, I. M. Ward, and J. E. McIntyre, Polymer 25:1291 (1984).
155. J. A. Siddiqui and P. V. Wright, Polymer Comm. 28:90 (1987).

156. M. Z. A. Munshi and B. B. Owens, Polymer J. (Tokyo) 20 (7):577 (1988).
157. M. Z. A. Munshi, B. B. Owens, and S. Nyuyen, Polymer J. (Tokyo) 20 (7):597 (1988).
158. T. K. Dzhumadilov, S. S. Ismagulova, and E. A. Bekturov, Izv. Akad, Nauk Kaz. SSR, SER. Khim. 1:87 (1989).
159. S. S. Ismagulova, T. K. Dzhymadilov, and E. A. Bekturov, Vysokomol. Soedin. Ser. B 31:209 (1989); C.A. 111: 7969 (1989).
160. M. A. Munshi and B. B. Owens, Gov. Rept. Announce. Index (USA), 88:19, Abst. No. 848,457 (1988); C.A. 110: 232407 (1989).
161. E. T. Lance-Gomez and T. C. Ward, J. Appl. Polymer Sci. 31:333 (1986).
162. M. Ataman, J. Macromol. Sci.-Chem. A24:967 (1987).
163. Y. L. Lee, Poly(ethylene oxide) Based Solid Electrolytes: Phase Diagram, Kinetics, and Ionic Conductivity, Dissertation, Northwestern University, Evanston, Illinois, 1987.
164. N. B. Graham, M. Zulfiqar, N. E. Nwachuku, and A. Rashid, Polymer 30:528 (1989).
165. P. Bloss, W. D. Hergeth, E. Doering, K. Witkowski, and S. Wartewig, Acta Polym. 40 (4):260 (1989).
166. S. A. Vshivkov, E. V. Rusinova, and M. A. Balashova, Vysokomol. Soedin. Ser. B. 30:847 (1988); C.A. 110: 58553 (1989).
167. I. N. Khan, D. Fish, Y. Delaviz, and J. Smid, Makromol. Chemie 190:1079 (1989).
168. A. Lebouc and J. C. Rabadeux, Eur. Polymer J. 24:603 (1988).
169. S. H. Jeon, S. M. Park, and T. Ree, J. Polym. Sci. Part A: Polym. Chem. 27:1721 (1989).
170. J. R. MacCallum and C. A. Vincent, eds., Polymer Electrolyte Reviews-1, Elsevier Applied Science, London, 1987.
171. J. Moacanin and E. F. Cuddihy, J. Polymer Sci. Part C 14:313 (1966).
172. R. E. Wetton, D. B. James, and W. Whiting, J. Polymer Sci. Polymer Ltrs. Ed. 14:577 (1976).
173. D. B. James, R. E. Wetton, and D. S. Brown, Polymer 20:187 (1979); Polymer Preprints, Am. Chem. Soc. Div. Polymer Chem. 19 (2):347 (1978).
174. J. R. Stevens and S. Schantz, Polymer 29:330 (1988).

175. R. Xue and C. A. Angell, Report TR-10, Order No. AD-A183502 (1987); C.A. 108: 205458 (1988).

176. S. S. Ismagulova, T. K. Dzhumadilov, and E. A. Bekturov, Izu, Akad. Nauk Kaz. SSR, Ser. Khim. 6:26 (1988); C.A. 110: 135873 (1989).

177. W. Ye and T. Mo, Ziraa Zashi 12 (1):78 (1989); C.A. 111: 8251 (1989).

178. S. Yoshikawa and D. J. Lyman, J. Polymer Sci. Polymer Ltrs. Div. 18:411 (1980).

179. J. F. LeNerst, A. Gandini, and H. Cheradame, Macromolecules 21:1117 (1988).

180. K. F. Wissbrun and M. J. Hannon, J. Polymer Sci. Polymer Phys. Ed. 13:223 (1975).

181. M. J. Hannon and K. F. Wissbrum, J. Polymer Sci. Polymer Phys. Ed. 13:113 (1975).

182. J. E. Glass, J. Phys. Chem. 72:4450 (1968).

183. J. E. Glass, J. Phys. Chem. 72:4459 (1968).

184. J. E. Glass, J. Polymer Sci. Part C, Polymer Symp. 34: 141 (1971).

185. J. E. Glass, R. D. Lundberg, and F. E. Bailey, Jr., Colloid Interface Sci. 33:491 (1970).

186. M. A. C. Stuart, J. T. F. Keurentjes, B. C. Bonekamp, and J. G. E. M. Fraaye, Colloid Surf. 17:91 (1986).

187. M. Kawaguchi, S. Komatsu, M. Matsuzumi, and A. Takahashi, J. Coll. Interface Sci. 102:356 (1984).

188. R. L. Shuler and W. A. Zisman, J. Phys. Chem. 74: 1523 (1970).

189. B. B. Sauer and H. Yu, Macromolecules 22:786 (1989).

190. J. Klein and P. F. Luckham, Macromolecules 17:1041 (1984).

191. H. Hommel, A. P. Legrand, H. Balard, and E. Papirer, Polymer 25:1297 (1984).

192. H. Hommel, A. P. Legrand, P. Tongne, H. Balard, and E. Papirir, Macromolecules 17:1578 (1984).

193. G. J. Howard, C. C. Ma, and C. W. Yip, Polymer Comm. 4:182 (1983).

194. M. Kawaguchi, A. Sakai, and A. Takahashi, Macromolecules 19:2952 (1986).

195. M. Kawaguchi, T. Hada, and A. Takahemshi, Macromolecules 22:4045 (1989).

196. D. J. Kuzmenka and S. Granick, Polymer Comm. 29:64 (1988).

197. M. Kawaguchi, M. Mikura, and A. Takahashi, Macromolecules 17:2063 (1984).

198. V. P. Yepifanov, Polymer Science USSR 20:1062 (1979); Vysokolmol. Soyed. A20 (4):942 (1978).
199. T. Cosgrove, T. G. Heath, and K. Ryan, Polymer Comm. 28:64 (1987).
200. J. W. Schurer, A. deBoer, and G. Challa, Polymer 16: 201 (1975).
201. E. Roerdink and G. Challa, Polymer 21:509 (1980).
202. E. J. Vorenkamp, G. Brinke, J. G. Mejer, H. Jager, and G. Challa, Polymer 26:1725 (1985).
203. V. S. Shaw, J. D. Keitz, D. R. Paul, and J. W. Barlow, J. Appl. Polymer Sci. 32:3863 (1986).
204. E. Martuscelli and G. B. Dema, in Polymer Blends: Processing Morphology and Properties (E. Martuscelli, R. Palumbo, and M. Kryszewski, eds.), Plenum Press, New York, 1980, p. 16.
205. E. Martuscelli, M. Canetti, L. Vincini, and A. Seves, Polymer 23:331 (1982).
206. E. Martuscelli, G. Demma, E. Rossi, and A. L. Segre, Polymer Comm. 24:266 (1983).
207. E. Martuscelli, M. Pracella, and W. P. Yue, Polymer 25: 1098 (1984).
208. E. Martuscelli, C. Silvestre, M. Canetti, C. DeLalla, and A. Seves, J. Poly. Sci. Part B: Poly. Phys. 27:1781 (1989).
209. M. M. Cortazar, M. E. Calahorra, and G. M. Gutzman, Eur. Polymer J. 18:165 (1982).
210. E. Calahorra, M. Cortazar, and G. M. Gutzman, Polymer 23:1322 (1982); Polymer 25:1097 (1984).
211. G. C. Alfonso, T. P. Russell, Macromolecules 19:1143 (1985).
212. H. Ito, T. P. Russell, and G. D. Wignall, Macromolecules 20:2213 (1987).
213. C. Silvestre, S. Cimmino, E. Martuscelli, F. E. Karasz, and W. J. MacKnight, Polymer 28:1190 (1987).
214. Y. Murakami, Polymer J. (Japan) 20:549 (1988).
215. S. Cimmino, E. Martuscelli, and C. Silvestre, Polymer 30: 393 (1989).
216. B. S. Kim and R. S. Porter, J. Polymer Sci. Part B: Polymer Phys. 26: 2499 (1988).
217. B. S. Kim and R. S. Borter, Macromolecules 18:1214 (1985).
218. Y. Zhao, B. Jasse, and L. Monnerie, Polymer 30:1643 (1989).

219. N. K. Kalfogulou, J. Polymer Sci.: Polymer Phys. Ed. 20:1259 (1982).
220. N. K. Kalfoglou, D. D. Sotiropoulou, and A. G. Margaritis, Eur. Polymer J. 24:389 (1988).
221. B. B. Sauer, M. Kawaguchi, and H. Yu, Macromolecules 20:2732 (1987).
222. B. B. Sauer, H. Yu, C.-F. Tien, and D. F. Hager, Macromolecules 20:393 (1986).
223. I. A. Katime, M. S. Anasagasti, M. C. Peleteiro, and R. Valenciano, Eur. Polymer J. 23:907 (1987).
224. B. T. Swinyard, J. A. Barrie, and D. J. Walsh, Polymer Comm. 28:331 (1987).
225. M. Avella and E. Martuscelli, Polymer 29: 1731 (1988).
226. K. Gonsalves and M. D. Rausch, J. Polymer Sci.: Polymer Chem Ed. 24:1419 (1986).
227. H. Younes and D. Cohn, Eur. Polymer J. 24:765 (1988).
228. X. Zhang, Huasue Tongbao 10:46 (1987); C.A. 108: 132592 (1988).
229. M. Hara and A. Eisenberg, Macromolecules 20:2160 (1987).
230. R. E. Wetton and P. J. Williams, Polymer Mater. Sci. England 59:846 (1988).
231. S. Cimmino, E. Martuscelli, C. Silvestre, M. Cannetti, C. DeLalla, and A. Seves, J. Poly. Sci. Part B: Poly. Phys. 27:1781 (1989).
232. D. Graebling, D. Froelich, and R. Muller, J. of Rheology 33:1283 (1989).
233. K. E. Min, J. S. Chiou, J. W. Barlow, and D. R. Paul, Polymer 28:1721 (1987).
234. M. Iriarte, J. I. Iribarrea, A. Etxeberria, and J. J. Iruin, Polymer 30:1160 (1989).
235. A. I. Bailey, B. K. Salem, D. J. Walsh, and A. Zeytountsian, Colloid Polymer Sci. 257:948 (1979).
236. M. Weber and R. Stadler, Polymer 29:1064 (1988).
237. M. Weber and R. Stadler, Polymer 29:1071 (1988).
238. J. Marchal and H. Benoit, J. polymer Sci. 23:223 (1957).
239. M. Davies, G. Williams, and G. D. Loveduck, Z. Electrochem. 64:575 (1960).
240. V. Magnasco, D. Dellepiane, and C. Rossi, Macromol. Chemie 65:16 (1963).
241. J. E. Mark and P. J. Flory, J. Am. Chem. Soc. 88:3702 (1966).
242. C. H. Porter and R. H. Boyd, Macromolecules 4:589 (1971).

243. T. M. Connor, B. E. Read, and G. Williams, J. Appl. Chem. (London) 14:74 (1964).
244. Y. Ishida, M. Matsuo, and T. Takayanagi, Polymer Ltr. 3:321 (1965).
245. A. J. Bur, Polymer 26:963 (1985).
246. M. E. Baur and W. H. Stockmayer, J. Chem. Phys. 43: 4319 (1965).
247. G. Williams, Trans. Faraday Soc. 61:1564 (1965).
248. W. H. Stockmayer, Pure Appl. Chem. 15:539 (1967).
249. S. Mashimo, S. Yagihara, and A. Chiba, Macromolecules 17:630 (1984).
250. P. E. Rouse, J. Chem. Phys. 21:1272 (1953).
251. B. H. Zimm, J. Chem. Phys. 24:269 (1956).
252. K. Pathmanathan, G. P. Johari, and R. K. Chan, Polymer 27:1907 (1986).
253. K. Pathmanathan and G. P. Johari, Polymer 29:303 (1988).
254. A. R. Blythe and G. M. Jeffs, J. Macromol. Sci.-Phys. B3:141 (1966).
255. J. Lal and G. S. Trick, J. Polymer Sci. Part A-1 8:2339 (1970).
256. J. A. Faucher, J. V. Koleske, E. R. Santee, J. J. Stratta, and C. W. Wilson, J. Appl. Phys. 37:3962 (1966).
257. B. E. Read, Polymer 3:529 (1962).
258. F. E. Bailey, Jr., and J. V. Koleske, in Poly(ethylene oxide), Academic Press, New York, 1976, pp. 136–140.
259. P. Tormala, Eur. Polymer J. 10:519 (1974).
260. E. Alfthan and A. deRuvo, Polymer 16:692 (1975).
261. H. Suzuki, J. Grebowicz, and B. Wunderlich, British Polymer J. 17:1 (1985).
262. N. G. McCrum, B. E. Read, and G. Williams, Anelastic and Dielectric Effects in Solids, John Wiley & Sons, New York, 1967.
263. J. A. Faucher, Polymer Ltrs. 3:143 (1965).
264. T. Alper, A. J. Barlow, and R. W. Gray, Polymer 17: 665 (1976).
265. J. M. G. Cowie, Polymer Eng. Sci. 19:709 (1979).
266. C. M. Wang, G. Fytas, D. Liege, and T. Dorfmuller, Macromolecules 14:1363 (1981).
267. G. P. Johari, A. Hallbrucker, and E. Mayer, J. Polymer Sci. Part B: Polymer Phys. 26:1923 (1988).
268. K. L. Smith and R. Van Cleve, Ind. Eng. Chem. 50:12 (1958).

269. B. S. Kim and R. S. Porter, _Macromolecules_ 18:1214
      (1985).
270. D. J. Mitchell and R. S. Porter, _Macromolecules_ 18:1218
      (1985).

7

Utility of the Poly(alkylene oxide)s

Although the utility of alkylene oxides and poly(alkylene oxide)s is mentioned at various points in this book, this chapter will discuss many of the use areas and detail some of the most important ones. In commercial use, the poly(alkylene oxide)s include principally the polyethylene glycols, poly(ethylene oxide), the polypropylene glycols, polyether polyols, random and block ethylene oxide/propylene oxide copolymers, and oxyethylene adducts of alcohols and phenols. Some butylene oxide is used as a comonomer with ethylene oxide and propylene oxide in polyether polyols and in polyethylene glycols. Propylene oxide and epichlorohydrin are used to make specialty elastomers (1). Otherwise, epichlorohydrin is used in the manufacture of epoxide resins for adhesives and the matrix of composites. There are some cyclic aliphatic epoxides as well, but again, they are largely in "epoxide" systems formulated especially for electrical insulation, electronic circuit boards, acid scavengers, and coatings.

By convention, the poly(alkylene oxide)s are linear or branched, soluble polymers and copolymers principally of ethylene oxide and propylene oxide. Again, by convention, poly-

ethylene glycols are low- to medium-molecular-weight polymers
of ethylene oxide, while polymers of ethylene oxide with molecular weights greater than about 50,000 are termed poly(ethylene
oxide) rather than glycols or polyols. The term polyoxyethylene is usually used with no implication of molecular weight.

Depending on the molecular weight, polymers of ethylene
oxide range from viscous liquids, molecular weights of 200 to
700; to semisolids, molecular weight of about 1000; to hard
waxes, molecular weights of 3000 to 20,000; to tough plastics
with molecular weights to several million. All these ethylene
oxide polymers are water soluble at room temperature, have
very low toxicity, are bland and nonirritating to the skin, have
wide compatibilities in formulations, and have good lubricity.
This unusual combination of properties has led the lower molecular weight polymers, the polyethylene glycols, to be widely
used in formulating pharmaceutical salves, ointments, suppositories, cosmetic creams and lotions, paper coatings, lubricants,
textile sizes, metal-working lubricants, and wood impregnants,
to name some of the diverse applications.

The largest single use of the polypropylene glycols and
polyols is in polyurethanes made by the reaction of the polyether polyols principally with tolylene diisocyanate (TDI) or one
of the polyisocyanates based on bis-(p-isocyanatophenylene)-
methane (MDI). These polyurethanes may be elastomers with a
broad range of hardnesses; injection-molded plastic parts; rigid
foam used for thermal insulation; or flexible polyurethane foams
used for upholstered furniture cushioning, seating, bedding,
carpet underlayment, packaging, and many other applications.
Many of these polyether polyols are copolymers containing a
smaller proportion of ethylene oxide and a major proportion of
propylene oxide, with the ethylene oxide often at the ends of
the molecules to provide a primary hydroxyl group with improved reactivity over the secondary hydroxyl group usually
found on propylene oxide polyols.

Random copolymers of ethylene oxide and propylene oxide
containing more than 40 to 50 percent ethylene oxide are water
soluble at room temperature. These liquid copolymers are used
as functional fluids and lubricants, hydraulic fluids, and metal-
working lubricants. Block copolymers—copolymers with
"blocks" of oxyethylene followed by "blocks" of oxypropylene
along the polymer chain—are soluble or dispersible in water and
are a valuable and unique class of surface-active agents widely
used as emulsion breakers of water-in-oil emulsions in the petroleum industry, as defoamers, and as low-sudsing detergents.

These copolymers were described in Chapter 4. A summary of industries that are among the principal users of poly(alkylene oxide)s is presented in Table 1.

In Chapter 6, molecular-association complexes of high-molecualr-weight poly(ethylene oxide)s were discussed. Table 2 contains an extensive list of compounds that have been reported to form molecular-association complexes that form the basis for some of the uses that have been developed (2).

I. SURFACTANTS

Archeochemists believe that the first surfactants, soaps, were made from animal fats and wood ashes in Sumeria about 2500 B.C. Soaps remained the detergents used through the millennia and were a relatively scarce item until the advent of modern chemistry in the nineteenth century. Today, the soap, surfactant, and detergent market occupies a major segment of the chemical industry. The U.S. market for detergents is about $11 billion per year, and world production of surfactants is about 15 billion pounds (6.75 million metric tons).

Detergents generally are formulated for specific purposes. Typical detergents contain up to 20 percent of an anionic or nonionic surfactant plus other formulation ingredients that have specific functional purposes, such as those detailed below.

Additive	Purpose
Soaps, silicon oils, paraffins	Suds-control agents
Fatty acid amides	Foam/suds stabilizers
Sodium tripolyphosphate	Chelators
Zeolite 4A	Ion exchanger
Sodium carbonate	Alkali
Sodium citrate	Cobuilder
Sodium perborate	Bleach
Quaternary ammonium compounds	Fabric softener
Cellulose ethers	Antiredeposition agent
Stilbene derivatives	Optical brighteners
Sodium silicate	Anticorrosion agent
Sodium sulfate decahydrate	Filler

TABLE 1   Industries That Use Poly(alkylene oxide)s

Industry	Applications
Agriculture	Water-soluble films as seed tapes and packaging, formulations of agricultural sprays
Automotive and transportation	Seating, interior padding, sound insulation, exterior molded parts
Ceramics	Binder, glass-fiber sizing, wet-flow and green-strength improvers
Construction	Adhesives, paving compositions
Dentistry	Adhesives, molding/modeling materials
Electrical and electronics	Battery separators, ionic conductors, antistatic materials
Electroplating	Metal grain size controllers, plate quality improvers
Food	Machinery and packaging lubricants, refrigeration insulation
Home and office furnishings	Seating, mattresses, carpet underlay, antistatic agents
Industrial supply	Rheology control agents, rubber and plastic processing
Metals and mining	Metal-working lubricants, hydraulic fluids, drilling lubricants, metal-cleaning formulations, flocculants, quenchants
Municipal services	Drag-reduction agents for fire fighting, sewage discharge controls
Paper	Filler-retention aids, white-water flocculation, surface improvers, adhesives, specialty additives
Personal-care products	Cosmetic lotions, lipsticks, chapsticks, ointments, skin lubricants, shaving creams and accessories

TABLE 1   (Continued)

Industry	Applications
Paints and finishes	Decorative and functional binders, surface cleaners and preparatives, flexibilizers for ultraviolet-light-cured varnishes and inks, moisture-cure and other polyurethanes, alkyds
Petroleum	Thickeners, defoamers
Pharmaceutical	Tablet and specialty coatings, pill binders, suppositories, enzyme modifiers, emollients
Printing and photography	Microencapsulated inks, photographic developer solutions, photographic emulsion stabilizers
Soaps and detergents	Emollients, thickeners, soil-suspending agents, anionic and nonionic surfactant synthesis
Textile	Dye assistants, antistatic agent additives, waste-water flocculants, sizings, lubricants, fabric heat-absorption modifiers, phase-transformation fabric

TABLE 2   Compounds Forming Association Complexes with High-Molecular-Weight Poly(ethylene oxide)   (From Ref. 2.)

Compound	Application
Bromine	Controlled release
Carboxymethyl dextran	Adhesives
Catechol tannins	Adhesives
Copolymer of methyl vinyl ether and maleic anhydride	Microencapsulation

TABLE 2   (Continued)

Compound	Application
Dichlorophene	Controlled release, pharmaceuticals
Gelatin	Microencapsulation
Hexachlorophene	Controlled release, pharmaceuticals
Iodine	Controlled release
Lignin sulfonates	Adhesives, emulsions
2-Naphthol	Adhesives
Lithium perchlorate	Ionic conductors
Phenolic resins, novolacs, and resoles	Solvent-soluble adhesives
Poly(acrylic acid)	Modification of poly(ethylene oxide) crystallinity, in analytical reagents
Polyureas	Modification of poly(ethylene oxide)
Potassium halides	Reduce crystallinity, increase antistatic properties
Potassium thiocyanate	Reduce crystallinity, ionic conduction
Sodium carboxymethyl-cellulose	Adhesives, microencapsulation
Thiolignin	Adhesives
Thiourea	Modify poly(ethylene oxide) crystallinity in formulations
Trimethylol phenol	With novolacs, hot-melt adhesives
Urea	Modify poly(ethylene oxide) in formulations, extend compatibility, in artificial kidneys

Specific detergent formulations vary widely in the United States
and are quite different in Europe and Japan with respect to the
relative amounts of the typical ingredients listed above.

About 45 percent of the world production of surfactants is
in the United States, with about 38 percent produced in Western
Europe and 17 percent in Japan. Of this production, about 65
percent is anionic and 28 percent is nonionic.

While the largest and most obvious use of surfactants is in
formulated detergents, there are many other applications, some
of which will be discussed in later parts of this chapter. Surfactants are used in many chemical processes, notably in water-
borne polymerization processes (3), textile manufacture, flow
and leveling of coatings, and leather processing. They are
also used in metal working and in lubricants; notably, in formulated automotive lubricants and oil.

Surfactant molecules are a unique class of chemical compounds that are made up of groups of opposite solubilities,
generally oil soluble and water soluble, and are designed to be
highly soluble in one of these phases. A characteristic of these
specialized molecules (and the reason they are termed surfactants) is that, at equilibrium, they are found in an enriched or
concentrated condition at the interface of two generally insoluble phases, rather than in the bulk of either phase. In addition, they are generally oriented at the phase interface. Above
a critical concentration in solution, surfactants aggregate into
micelles.

Surfactants fall into four major categories: anionics, cat-
ionics, nonionics, and amphoterics. The poly(alkylene oxide)s
are used in anionic surfactants as alcohol polyether sulfates,

$$R-\overset{\overset{\textstyle R'}{\textstyle |}}{C}H-CH_2-O-(CH_2-CH_2-O-)_n-SO_3Na \qquad \begin{array}{l} n = 1-4 \\ R' = \text{alkyl or} \\ \text{hydrogen} \end{array}$$

and in nonionic surfactants as alkylphenolethyoxylates,

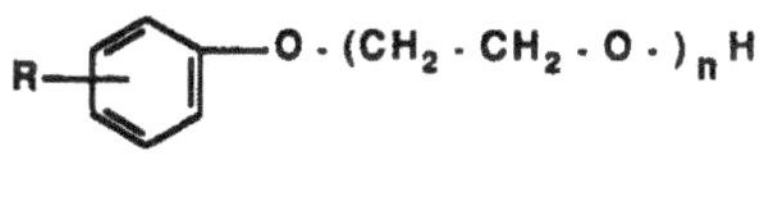

$$R-\langle\!\!\!\bigcirc\!\!\!\rangle-O\cdot(CH_2\cdot CH_2\cdot O\cdot)_n H$$

$$n = 5-10$$

as alcohol polyethylene glycol ethers,

$$R-\overset{\overset{\displaystyle R'}{|}}{C}H-CH_2-O-(CH_2-CH_2-O)_nH$$

$n = 3-15$
$R' =$ alkyl or hydrogen

as ethylene oxide/propylene oxide block copolymers,

$$H(O-CH_2-CH_2)_n-O-(CH_2-CH(CH_3)-O)_mH$$

$n = 2-60$
$m = 15-80$

and as fatty alcohol polyglycol ethers,

$$RO-(CH_2-CH_2-O)_n(CH_2-CH(CH_3)-O)_mH$$

$n = 3-6$
$m = 3-6$

The first poly(alkylene oxide)-derived surfactants were patented by BASF in 1930. Since then, an enormous growth in manufacture and use has occurred, along with a huge body of scientific and engineering information (4).

Some of the uses of the poly(alkylene oxide)s in surfactants will also be noted in connection with specific applications in later portions of this chapter. A few specific applications, however, are listed here. Ethoxylated fatty alcohol nonionic surfactants are used as a low-froth dye auxiliary in textiles (5). Polyethylene glycols are used as emulsifiers and as spreader-stickers for water-borne agricultural herbicides and insecticides. Polyethylene glycols, while not themselves detergents, are formulated into detergents as soil-suspending agents. Alteration of the surface properties of paper without adversely affecting the water-retention properties and behavior of optical brightness is accomplished by spread-cotaing compositions that include from 5 to 20 percent of polyoxyethylene that has a molecular weight of 6000 to 50,000 (6). Poly(ethylene oxide) with a molecular weight of about 400,000 is added at about 0.1 percent to liquid detergents for hand use to give the skin a smoother, supple feel (7). Specialty surfactants that contain disulfide groups have been prepared by propoxylation and ethoxylation of dihydroxyl functional disulfides (8).

II. LUBRICANTS

The poly(alkylene oxide)s have been widely used in formulating synthetic lubricants. Their use has been based on several

unique properties of this class of materials. First, the poly-
alkylene oxides—essentially polyethylene glycols, polypropylene
glycols, and copolymer glycols—have been readily available
commercially as well-defined and well-characterized molecular
species. These liquids are available with narrow molecular
weight distributions, which has meant that they do not have low-
molecular-weight, volatile fractions. The polymers have unique-
ly low thermal coefficients of viscosity, or viscosity index. The
viscosities are relatively constant over a use temperature
range—that is, they do not thin-out or exhibit marked viscosity decreases as the temperature rises in service. They have
wide compatibilities and solubilities and are easily formulated in
both aqueous and oleophilic systems. The polymers inherently
wet many surfaces ranging from metals to natural and synthetic
textiles. Their toxicity is low. The decomposition products of
the polyglycols are volatile and, on pyrolysis, do not leave
char or carbonaceous residues (see Chapter 4, 5, and 6).

Because of their water solubility and low toxicity, polyethylene glycols are used as lubricants in food-handling machinery
and equipment that may contact food products. Polyethylene
glycols are used as textile lubricants where the hydrophilic
properties also impart antistatic properties during processing.
Low toxicity is also the reason that polyethylene glycols are
used as lubricants for surgical sutures.

Polypropylene glycols were used in a limited way in the
mid-1950s and were among the first synthetic lubricants to be
used as automotive engine oils. Performance was excellent;
however, incompatibility with the most widely used hydrocarbon
oils was a problem. The main features of these automotive oils
were their wide temperature service range—i.e., suitable for
northern winters as well as Gulf Coast summers—and volatile
decomposition products that left no char and a clean engine.
However, the advent of multigraded engine oils and detergent
oil formulations proved to have market advantages over the
more expensive poly(alkylene oxide) automotive lubricants.

The same features, however, proved successful in the commercial use of the poly(alkylene oxide)s as machine lubricants,
metal-working lubricants, quenchants, and hydraulic fluids.
The solubility of ethylene oxide/propylene oxide copolymer glycols in water has led these formulations to be universally accepted as "hydrolubes" and brake fluids where the nonflamma-
bility and wide temperature service range of a blend of polyol
and water are key features.

A very interesting use of the lubricant properties of a
poly(alkylene oxide) is a water/surfactant/isopropanol solution

containing about a half percent of poly(ethylene oxide) of 4 million molecular weight. This solution is used as a tire-mounting lubricant to ease the mounting of tubeless automobile tires on the wheel rim (9).

III. PHARMACEUTICALS AND MEDICINE

Because of low toxicity and good water solubility, the polyethylene glycols and poly(ethylene oxide) waxes have found wide use in pharmaceutical formulations and in the biochemical laboratory. In pharmaceutical use, there is generally a specification that limits usage to polymers above certain low molecular weight limits. However, because of the polymerization characteristics of ethylene oxide, which give polymers of narrow molecular weight distribution, these limits are easily met. A method for manufacture of polyethylene glycols for pharmaceuticals has been described (10). Propylene glycol is also used.

Polyethylene glycols, particularly as semisolids and waxes, are used in formulating suppositories and ointments, because the human body is highly tolerant of these polyglycols. Combinations with propylene glycol are also used. The desired melting point can be adjusted by choosing the polyglycol molecular weight. Xylocaine^{TM} is an example of an ointment that contains 5 percent lidocaine, a substituted acetamide that is the active ingredient, and 95 percent a mixture of polyethyleneglycol 1500, polyethyleneglycol 4000, and propylene glycol. Crystalline polyethylene glycols are also used in formulating medicinal tablets to make the tablets stronger and more chip resistant.

Polyethylene glycols are used in both humans and animals as carriers of drugs for intramuscular injection. Polyethylene glycols are also used as dialysis agents to remove water from protein and blood.

In the pathology and biochemical laboratories, polyethylene glycols are used to embed histological samples and pathology specimens. Specimens are sometimes preserved in polyethylene glycol and imbedded in solid polyethylene glycol such as the 4000-molecular-weight waxes. Because of the unique complexing ability of the poly(ethylene oxide) chain in water solution to form ionic complexes with salts, polyethylene glycols are used in manipulating cells. The polyether aids in the penetration of the cell membranes without damage, while not having the devastating effect of the crown-ethers on cell transpiration. Ureaabsorbent polymeric microspheres with a polyoxyethylene glycol derivative inner layer have a high urea-adsorbing power and can be regenerated by water washing, making them useful in artificial kidneys (11).

Microspheres also have been prepared from poly(methacrylic acid-g-ethylene oxide) (11a). The ability of the copolymer to form reversible hydrogen-bonded complexes suggests a potential for use as biosensors and/or carriers for controlled drug delivery. In a physiological environment, the copolymer was more useful in a microsphere form than in the usual helical or random-coil form.

High-molecular-weight poly(ethylene oxide) has been used because of its dispersant and lubricating characteristics. High-molecular-weight poly(ethylene oxide) is used as a dispersant for calamine lotion to prevent settling (12). Highly water-absorbent hydrogels are made by crosslinking, either chemically or with high-energy electron radiation, high-molecular-weight poly(ethylene oxide) in water solution in the concentration range of about 5 percent (13). These hydrogels have been used for culture media, replacing agar.

About 1/10 percent of a 4-million-molecular-weight poly(ethylene oxide) in rubbing alcohol gives a compound that is much improved for massage uses. The compound spreads more easily, with more lubricity, and leaves the skin with a smoother feel. High-molecular-weight poly(ethylene oxide) is used in denture fixative formulations, along with a natural gum such as gum karaya, to increase wet-tack to give a superior denture adhesive (14). Addition of crystalline poly(ethylene oxide) to a proprietary, reactive bone-cement powder decreased the polymerization temperature rise occurring during adhesive cure from 108°C to 65°C and thus markedly reduced the thermal trauma suffered by surrounding tissue (15).

Enzyme drugs have been coated with polyethylene glycols to mask the enzyme and prevent white blood cells from rapidly destroying the foreign material (16). Such masked drugs last longer and have increased effectiveness. Chapter 5, Section V, also contains a discussion about the use of radiation-cross-linked poly(ethylene oxide) for controlled drug-delivery systems as well as for blood filtration.

Crosslinked blends of poly(ethylene oxide) and poly(propylene oxide) were studied (17) to elucidate the blood compatibility exhibited by polyurethanes that have been made with polyethylene glycols and polypropylene glycols as their soft segments

(18, 19). The results indicate that the good compatibility may be related to an amorphous, poly(ethylene oxide)-enriched, hydrophilic surface that is presented to the blood.

IV. POLYURETHANE INTERMEDIATES

A. Elastomers

In 1937, Otto Bayer at the Leverkusen research laboratories of Bayer, AG, discovered the first of the many useful products that can be derived from the reaction of polyisocyanates with glycols (or polyols) and with diamines. The reaction of diisocyanate with glycol gave polyurethanes

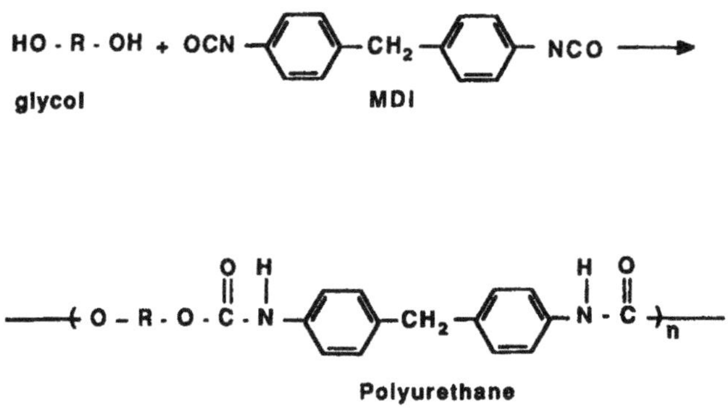

glycol   MDI

Polyurethane

and the reaction of diisocyanate with diamine gave a polyurea.

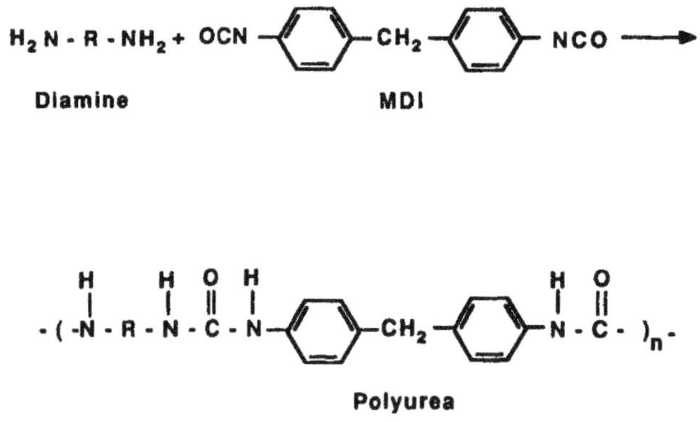

Diamine   MDI

Polyurea

Today, polyurethanes, polyureas, and urea/urethane segmented block copolymers find use as elastomers, fibers, coatings, printing rolls, adhesives, sealants, rigid foams for thermal insulation and flexible foams for home and office furnishings,
automobile seating and sound insulation, carpet underlay, and
packaging.

Among the earliest of the polyurethane products were elastomers and fibers. One fiber marketed as PERLON^{TM} was
manufactured by Bayer and was the product of 1,4-butanediol
and hexamethylene diisocyanate

$$HO-CH_2-CH_2-CH_2-CH_2-OH \ + \ OCN-(CH_2)_6-NCO \ \longrightarrow$$

1,4-butanediol   hexamethylene
                                         diisocyanate

$$-\left(O-(CH_2)_4-O-\overset{\overset{O}{\|}}{C}-\overset{\overset{H}{|}}{N}-(CH_2)_6-\overset{\overset{H}{|}}{N}-\overset{\overset{O}{\|}}{C}\right)_n-$$

polyurethane for fiber

Elastomers can be formed by using higher molecular weight
diols, particularly diols that would be amorphous, rubbery, and
flexible structural units in the linear polyurethane chain. For
this use, poly(propylene oxide) glycols and polyols are particularly well suited.

The synthesis of the poly(alkylene oxide)s for elastomers
is contained in Chapter 4, Sections III.A, IV, and V. For
polyurethane elastomers based on propylene oxide, the polyethers prepared are secondary hydroxyl-terminated polyethers.
The secondary hydroxyl group is slower to react with isocyanate than is a primary hydroxyl (Table 3) (20). Therefore,
poly(propylene oxide) diols and higher functionality polyols are
often capped with ethylene oxide (see Chapter 4, Sections IV
and V) to obtain the more reactive terminal, primary hydroxyl
groups.

The poly(alkylene oxide)s for urethane applications are
generally characterized by composition, molecular weight, and
functionality (also see Chapter 4, Section V, and Chapter 5,
Section I.A. Molecular weight and functionality are related by
hydroxyl number. Hydroxyl number is determined by titration
of the fully reacted phthalic anhydride adduct of the monol,
diol, or polyol in question and is expressed in terms of milli-

TABLE 3   Qualitative Reactivities with Isocyanates (From Ref. 20.)

	Relative reaction rate
Hydroxyls with phenylisocyanate	
Primary OH	1.0
Secondary OH	0.3
Tertiary OH	0.005
Isocyanates with n-butanol	
2,4-tolylene diisocyanate	
4-NCO	1.0
2-NCO	0.2
4,4'-diisocyanatodiphenylmethane (MDI)	0.75

grams of potassium hydroxide required to neutralize the mono-phthalate formed (21). Functionality, which is the average number of hydroxyls per polyol molecule, is nominally the functionality of the starter molecule used in preparing the polyol (e.g., glycol would be 2, glycerol would be 3, sorbitol would be 6, etc.).

For elastomers, tolylene diisocyanate (TDI) and 4,4'-diphenylmethane diisocyanate (MDI) have come to be the principal multifunctional isocyanates used in polyurethane and polyurea manufacture. Tolylene diisocyanate is usually obtained as a commercial 80/20 by weight percent mixture of the 2,4- and the 2,6- isomers. A 65/35 mixture of these isomers is commercially available, as is the pure 2,4- isomer. The 2,4- isomer is used in selected coating applications and in the making of certain prepolymers. The 65/35 mixture of isomers is a result of separation of pure 2,4- isomer from the 80/20 mixed product of normal manufacture.

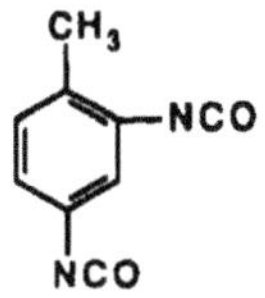

2, 4-Tolylene
diisocyanate
80%

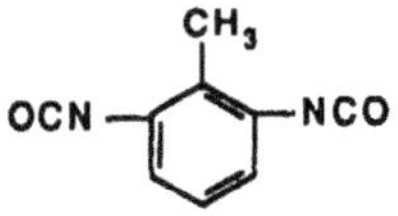

2, 6-Tolylene
diisocyanate
20%

MDI is available commercially in several forms. The pure 4,4'-diisocyanatodiphenylmethane is a crystalline solid at room temperature and is used only for fiber and some very special elastomer products. There are forms of MDI that are liquid at room temperature and have a slightly higher functionality, between 2 and 3 (22). A typical liquid MDI is a mixture of MDI and its dimer and trimer and can be represented as follows:

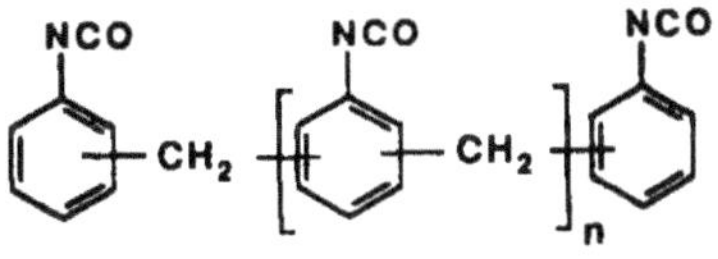

Liquid MDI

An elastomer is formed by the reaction of a higher molecular weight diol—i.e., molecular weights of about 500 to 2000—with a diisocyanate.

$$
HO-R-OH + OCN-R'-NCO \rightarrow \left[O-R-O-\underset{\underset{\displaystyle \text{}}{}}{\overset{\overset{\displaystyle O}{\|}}{C}}-\underset{\underset{\displaystyle \text{}}{}}{\overset{\overset{\displaystyle H}{|}}{N}}-R'-\underset{\underset{\displaystyle \text{}}{}}{\overset{\overset{\displaystyle H}{|}}{N}}-\underset{\underset{\displaystyle \text{}}{}}{\overset{\overset{\displaystyle O}{\|}}{C}} \right]_n
$$

diol   diisocyanate   polyurethane

When an excess of diisocyanate is used, crosslinking occurs through formation of an allophanate, which is the reaction product of an already formed urethane group and an isocyanate

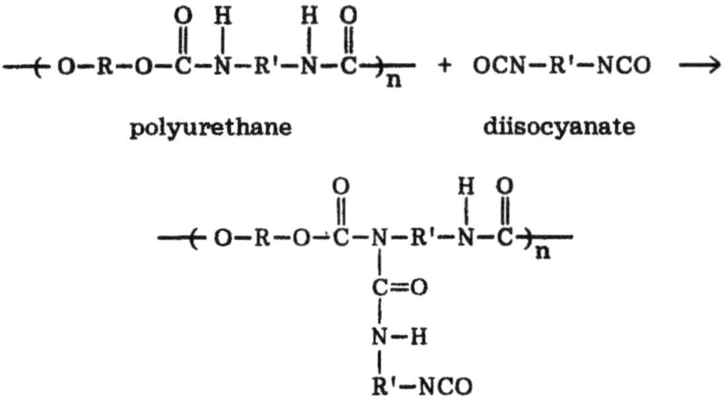

$$\begin{array}{c} \quad\quad\;\; \overset{\text{O}}{\overset{\|}{}} \; \overset{\text{H}}{\overset{|}{}} \quad\quad \overset{\text{H}}{\overset{|}{}} \; \overset{\text{O}}{\overset{\|}{}} \\ -\!\!\!-\!\!\!(\,\text{O-R-O-C-N-R'-N-C}\,)_{\!n} \;+\; \text{OCN-R'-NCO} \;\longrightarrow \end{array}$$

polyurethane   diisocyanate

$$\begin{array}{c} \quad\quad\quad\quad\quad \overset{\text{O}}{\overset{\|}{}} \quad\quad \overset{\text{H}}{\overset{|}{}} \; \overset{\text{O}}{\overset{\|}{}} \\ -\!\!\!-\!\!\!(\,\text{O-R-O-C-N-R'-N-C}\,)_{\!n} \\ \quad\quad\quad\quad\quad\; \overset{|}{\underset{}{}} \\ \quad\quad\quad\quad\quad\; \text{C=O} \\ \quad\quad\quad\quad\quad\; \overset{|}{\underset{}{}} \\ \quad\quad\quad\quad\quad\; \text{N-H} \\ \quad\quad\quad\quad\quad\; \overset{|}{\underset{}{}} \\ \quad\quad\quad\quad\quad\; \text{R'-NCO} \end{array}$$

isocyanate-terminated allophanate

The terminal isocyanate group can react with polyol molecules—
i.e., hydroxyl groups—or with other formed urethane groups,
and crossliking results.

A particularly useful method of formulating polyurethane
elastomers based on poly(propylene oxide) diols uses a prepolymer of a diisocyanate and the diol. The prepolymer polyether,
usually a poly(propylene oxide) diisocyanate, has a molecular
weight of about 1000 to 2000.

$$\text{HO-}(CH(CH_3)-CH_2-O)_n-O-CH_2CH(CH_3)-O-(CH_2-CH(CH_3)-O-)_mH \quad +$$

Poly(propylene oxide) diol

$$\text{2 OCN-R-NCO} \longrightarrow$$

Diisocyanate

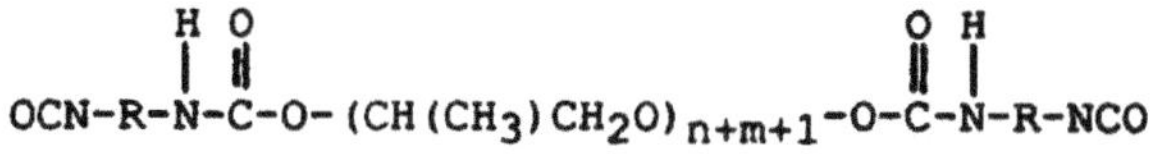

$$\overset{\text{H}\;\;\text{O}}{\underset{}{\text{OCN-R-N-C-O-}(CH(CH_3)CH_2O)_{n+m+1}\text{-O-C-N-R-NCO}}}$$

Linear isocyanate-terminated prepolymer

The prepolymer can be cured with a chain extender, a diol or triol, or water, in a moisture-cured system. The chain extender can be a triol such as glycerin or a tetrol such as pentaerythritol, as well as similar compounds.

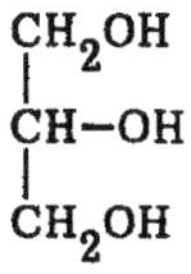

$$CH_2OH$$
$$|$$
$$CH-OH$$
$$|$$
$$CH_2OH$$

glycerine

$$
\begin{array}{ccc}
HOCH_2 & & CH_2OH \\
 & \diagdown \quad \diagup & \\
 & C & \\
 & \diagup \quad \diagdown & \\
HOCH_2 & & CH_2OH \\
\end{array}
$$

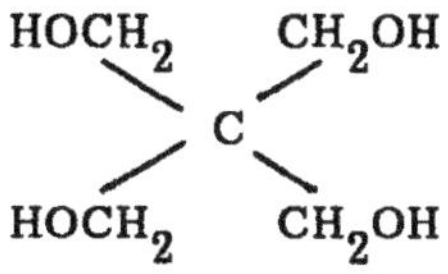

pentaerythritol

Prepolymers can be cured by an excess of diisocyanate or by a combination of triols and polyols with prepolymer. The formation of urethane linkages by reaction of hydroxyl groups with isocyanate groups is catalyzed by a number of organometallic compounds. Most commonly, tin compounds such as stannous octoate or dibutyltindilaurate are used. Many other metal compounds have been used, including those of cobalt, lead, and copper (23).

If a small amount of water is added to the polyether isocyanate prepolymer as a chain extender, the reactions that occur will introduce urea groups into the polymer chain, because isocyanate groups react with water to yield an amine and liberate carbon dioxide.

$$OCN-R-NCO \; + \; H_2O \; \rightarrow \; OCN-R-NH_2 \; + \; CO_2$$

The amine group then reacts with an isocyanate group to produce a urea linkage that has improved association characteristics over those of the urethane linkages, and improved mechanical properties result.

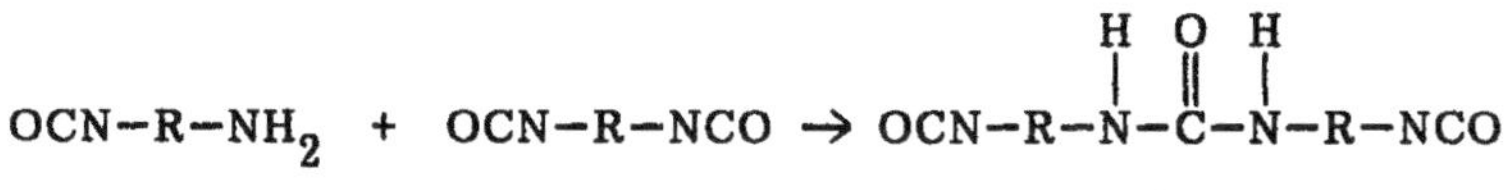

$$
OCN-R-NH_2 \; + \; OCN-R-NCO \rightarrow OCN-R-\overset{\displaystyle H}{\underset{\displaystyle |}{N}}-\overset{\displaystyle O}{\underset{\displaystyle \|}{C}}-\overset{\displaystyle H}{\underset{\displaystyle |}{N}}-R-NCO
$$

urea linkage

The urea-linked prepolymer segments can then react with additional isocyanate groups and form biurets.

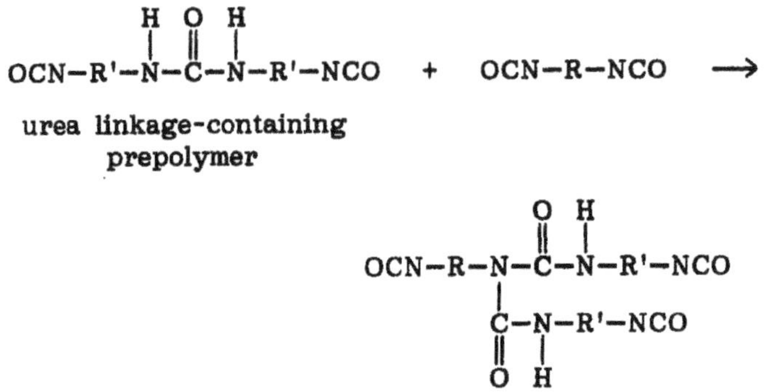

$$OCN-R'-N-\overset{\overset{\displaystyle H}{|}}{\underset{\underset{\displaystyle H}{|}}{N}}-\overset{\overset{\displaystyle O}{\|}}{C}-N-R'-NCO \quad + \quad OCN-R-NCO \quad \longrightarrow$$

urea linkage-containing
prepolymer

$$OCN-R-N-\overset{\overset{\displaystyle O}{\|}}{C}-\overset{\overset{\displaystyle H}{|}}{N}-R'-NCO$$

$$\underset{\underset{\displaystyle O \quad H}{\| \quad |}}{|}$$

$$C-N-R'-NCO$$

biuret-polyisocyanate
prepolymer

Isocyanate-terminated urethane prepolymers can be formed and cured into an elastomer by casting the prepolymer and allowing ambient moisture to react with the cast system's end groups. Moisture-cured sealants can be formulated and allowed to cure in a similar manner.

B. Reactive Processing (RIM)

Reactive processing of systems formulated to produce urethane elastomers, or reaction injection molding (RIM), is a technology that was developed to produce hard, tough, elastomeric, shaped objects. The initial incentive for this development was the Motor Vehicle Safety Standard 215 regulation in the United States, which required that the front and rear ends of an automobile be able to withstand a 5-mph impact without impairing functional parts of the car. To meet this requirement, it was necessary to completely alter chromed metal bumpers and to design a new class of bumpers. Automotive engineers turned to elastomeric materials containing urethane and urea linkages to meet the requirement, because such polymers had high quality and the capability of very rapid, but controllable, reactivity. Since that initial impetus, more extensive applications for RIM technology have developed.

RIM elastomers are also termed microcellular elastomers. The molded parts produced by RIM processing consist of a microcellular core encapsulated by an integral skin of higher density elastomer of the same chemical composition (24).

Reaction injection molding has outward similarities to thermoplastic injection molding (TIM), but is fundamentally differ-

ent. The major difference is that in TIM, a previously synthesized polymer is heated, liquefied, and injected into a mold
cavity. It is then cooled, and the cooling of the melt is used
to give the solid polymer and the desired shape of molded product. In contrast, RIM uses monomeric species such as multifunctional isocyanates and polyols that are rapidly injected into
the mold cavity. Rapid polymerization takes place, and solid
polymer forms in the mold as a result of polymerization and
crosslinking. With certain systems, phase separation of the
formed species occurs during and after the rapid reaction.
With RIM, the mold-demold time, which represents the time at
which liquid raw materials are converted into solid molded objects, can be very short, around 1 to 3 minutes (25).

In molding processes such as thermoset injection molding
(for example, with phenolics), a hot mold is used to initiate
the polymerization/crosslinking reactions. In RIM, mold temperature does not play a key role. The reaction is initiated in an
intensive mixing process that blends the reactive chemicals just
before and as the reaction mixture is injected into the mold.

Early in the development of polyurethane elastomers and
foams, which will be discussed in the next section, both elastomers and foams were made by mixing the reactive components—
basically diisocyanate and the formulated polyol—by low-pres-
sure injection through rotating mixers. High-pressure impinge-
ment mixing was developed to give more rapid and more intimate
mixing (26).

The first major growth of RIM processing, as noted earlier
in this section, occurred in the United States with the development of automobile bumpers designed to meet Congressionally
mandated safety standards in a 5-mph impact. Polyurethane
fascia (a term applied to both the front and rear bumpers, as
well as fenders and other automobile parts) were designed to
cover steel beams mounted on shock absorbers. Later, glass-
reinforced RIM (RRIM) products were introduced and applied to
some fender and body parts. In 1987, about 150 million pounds
(67,500 metric tons) of material were used for the manufacture
of RIM products in the United States. About half of this total
is polyether polyol, and about half was used in automobiles.

RIM formulations are generally in two parts; basically, the
polyisocyanate and the polyol. In RIM, the predominantly used
isocyanates are derivatives of MDI. MDI is preferred over TDI
because its lower vapor pressure makes handling less dangerous
and because the product physical properties are usually superior. Since MDI is a solid at ambient temperatures, the liquid

MDIs are preferred. The preferred liquid MDIs are principally the 4,4'-diphenylmethane diisocyanate containing some dimer or polymeric species to reduce the melting point below normal handling temperatures (22).

The polyols of greatest use are poly(alkylene oxide)s, almost exclusively poly(propylene oxide) that is usually capped with ethylene oxide to produce a polyol with terminal, primary hydroxyl groups. Molecular weights of these polyols are usually in the range of 200 to 6000 and functionalities are 2 to 4. Poly(propylene oxide) is preferred because of its low glass transition temperature, $T_g = -60^{\circ}C$ (27), which results in polyurethanes with good low-temperature characteristics and elastomeric properties over a broad and useful temperature range. It is preferred over poly(ethylene oxide) because of its much lower water sensitivity and over polyester and polycaprolactone polyols because of cost and hydrolytic stability factors. The poly(propylene oxide) polyols have relatively lower viscosities for a given molecular weight than do the corresponding polybutadiene or polyisobutylene polyols and also have superior viscosity indexes (lower temperature coefficients of viscosity) in comparison with the hydrocarbon polyols.

In general, in formulating RIM products, <u>chain extenders</u>, or low-molecular-weight diols, are used to control product mechanical properties, especially stiffness modulus. The lower the molecular weight of the chain extender, the greater the product stiffness modulus. Extenders that can be used include ethylene glycol, 1,4-butanediol, and certain amines.

RIM products are best understood from a structure-property standpoint as segmented block copolymers. Polymers with such structures can be schematically described and compared in the following manner.

1. An ideal elastomer is a network of amorphous segments crosslinked at specific sites along the polymer chain. This network obeys the classic theory of rubber elasticity as described by Flory (28) and may be represented as follows.

The network of an ideal elastomer has a Poisson's ratio of 0.5 and returns completely to its initial state after deformation, with no loss of energy as heat resulting from the deformation and recovery. Of course, real elastomers are designed to approximate this ideal behavior as well as possible.

2. A polyurethane network differs from the above representation; it can be described by the following schematic representation.

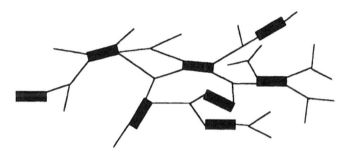

In this representation, crosslinks are pictured as

These crosslinks are derived through the functionality of the polyol, which may be a triol or a tetrol, for example. The other crosslinks, which can be represented as

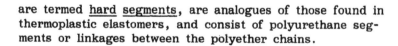

are termed <u>hard</u> <u>segments</u>, are analogues of those found in thermoplastic elastomers, and consist of polyurethane segments or linkages between the polyether chains.

$$R\text{-}O\text{-}\underset{\underset{O}{\|}}{C}\text{-}\underset{\underset{H}{|}}{N}-\!\!\!\bigcirc\!\!\!-CH_2-\!\!\!\bigcirc\!\!\!-\underset{\underset{H}{|}}{N}\text{-}\underset{\underset{O}{\|}}{C}\text{-}O\text{-}R'$$

These more polar, rigid polyurethane linkages can phase-separate and aggregate, to some degree, particularly if these units are chain-extended by short chain diols such as ethylene glycol or 1,4-butanediol.

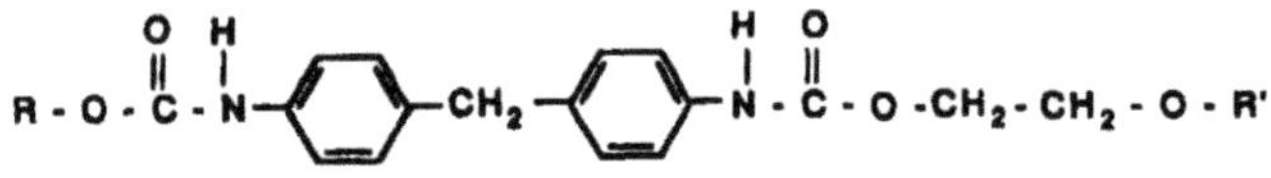

MDI chain extended with ethylene glycol

These secondary or pseudo crosslinks, within the use temperature range of the product, increase the stiffness modulus. As indicated above, this type of crosslinking is achieved by phase separation or aggregation and forms over a period of time that ranges from a few hours to several days, depending on storage conditions. Mechanical properties can change over this time period, with improvements usually obtained.

3. In recent years, aromatic amine chain extenders have been introduced that have largely replaced the diols as chain extenders. The most widely used amine is an 80/20 mixture of two isomers: 2,4- and 2,6-diamino-3,5-diethyltoluene.

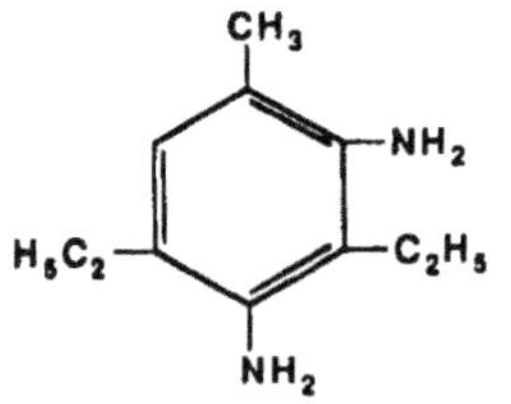

2, 4-Isomer
80%

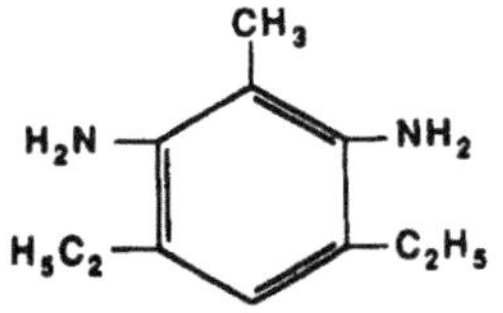

2, 6-Isomer
20%

The 4-position isomer is the least reactive group, and in the 2,6-diamine isomer, the second amino group to react has about one-third the reactivity of the first amino group to react (29–32). Use of this diamine chain extender results in substituted polyurea units in the chain.

$$\text{H}_5\text{C}_2 \overbrace{}^{\text{CH}_3} \text{(ring, H}_2\text{N, C}_2\text{H}_5)\text{-N}\overset{\text{H}}{\underset{}{}}\text{-}\overset{\text{O}}{\underset{}{\text{C}}}\text{-N}\overset{\text{H}}{\underset{}{}}\text{-}\bigcirc\text{-CH}_2\text{-}\bigcirc\text{-N}\overset{\text{H}}{}\text{-}\overset{\text{O}}{\text{C}}\text{-N}\overset{\text{H}}{}\text{-(ring, CH}_3, \text{NH}_2, \text{C}_2\text{H}_5)\text{-C}_2\text{H}_5$$

These urea groups are higher softening, are more polar, and phase-separate more completely than urethane groups obtained from diol chain extenders and, therefore, produce a stiffer, more highly reinforced, higher service temperature RIM product than a urethane-linked product.

4. The most recent modification in RIM formulations based on poly(alkylene oxide) elastomeric components has been amine-terminated poly(alkylene oxide) polyethers (see Chapter 5, Section III).

$$\text{H}_2\text{N-}\underset{\text{CH}_3}{\text{CH}}\text{-CH}_2\text{-O-}(\underset{\text{CH}_3}{\text{CH}}\text{-CH}_2\text{-O})_m\text{-CH}_2\text{-}\underset{\text{CH}_3}{\text{CH}}\text{-O-}(\text{CH}_2\text{-}\underset{\text{CH}_3}{\text{CH}}\text{-O})_n\text{-CH}_2\text{-}\underset{\text{CH}_3}{\text{CH}}\text{-NH}_2$$

These amine-terminated poly(alkylene oxide)s, which have amine functionalities of 2 or 3, are produced by Texaco Chemical. They are extremely fast-reacting with isocyanates and produce polyurea hard segments in the elastomer network (33, 34).

Microcellular RIM products are achieved by addition of blowing agents, which may be small amounts of fluorocarbon, to produce small gas bubbles in the elastomer product. It is also possible to disperse small, controlled amounts of air or nitrogen during high-intensity mixing to provide the microcellular structure. The expansion of the small gas cells in the elastomer during curing of the RIM product has two effects. One effect is to reduce the product density, and the other is to compensate for the shrinkage (molar volume reduction) during polymerization in the mold by expansion of the small gas cells.

The complete formulation of the RIM product includes polyol, isocyanate, chain extender, dispersed gas or blowing agent, surfactant to help control the gas cell dispersion in the elastomer, tin catalyst to accelerate the polyurethane formation (not needed with the amine-terminated polyethers), and fillers that may be of the reinforcing type. The most widely used rein-

forcing fillers are polymer polyols, which were discussed in
Chapter 5. These polymeric fillers are usually either polyacrylonitrile or styrene-acrylonitrile copolymers dispersed as fine
particles in the polyol and are prepared by graft polymerization
in situ in the polyol. Polymer polyols are also discussed in the
next section of this chapter, which deals with their use in urethane foams.

A schematic representation of the RIM molding process is
shown in Figure 1. More detail on RIM processing can be
found in several references (23, 35, 36).

The principal advantages of RIM processing of polyurethanes
and polyureas are the ability to mold large parts such as automobile fascias and body panels, motor housings, and bodies of
recreational vehicles, and that the molded parts made can be
lightweight. The microcellular structure of RIM parts permits
densities to be quite low, while the stiffness modulus can be

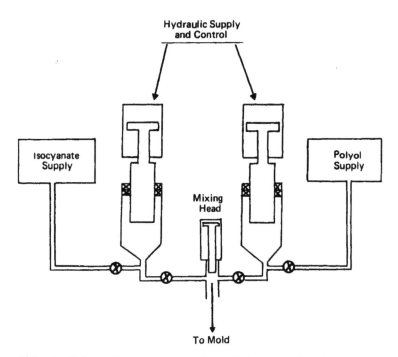

FIG. 1   Schematic representation of the reaction injection
molding (RIM) process.

varied over a wide range by adjusting the polyol molecular
weight—and the degree of crosslinking through polyol, amine,
and/or isocyanate functionality—by selecting among polyurethane, polyurethane/urea, and polyurea matrices and by opting
for or against a reinforcing filler, which may be polymer polyol
or a fiberglass.

While these material factors allow a wide range of stiffness
moduli and other mechanical characteristics to be designed into
molded objects, the stiffness of the molded part itself can also
be modified and designed by varying molded part thickness
and/or using ribs or beams within the molded part. For a
given molded part surface area, the weight of the part varies
with the thickness of the molding; however, the stiffness varies
with the cube of the thickness.

Part area $= xy$

Part weight $\propto r$

Part stiffness $\propto r^3$

Thus, very stiff moldings can be obtained by increasing the
molded part thickness without large increases in the weight of
the molded part.

C. Polyurethane Foams

Polyurethane foams are complex chemical and physical structures
that can be thought of as highly dilute composites. These foams
consist of two phases. One phase is a solid, polymeric, continuous phase; the second is a gaseous phase that may be either discontinuous (a closed-cell foam) or continuous (an open-
cell foam). A further distinction is made between rigid foams
and flexible foams (20, 36).

Rigid, or structural, foams are used principally for their
thermal insulation properties and are, therefore, almost always
closed-cell foams. Flexible foams are used principally as seating, cushioning, carpet underlay, fabric backing, and packaging and are generally open-cell foams. Some flexible foams
used in special applications such as sports equipment—for example in backpacking, where the foam may contact the ground
or be exposed to water and the sponge effect of an open-cell
foam would be undesirable—may be closed-cell foam products.

In any case, the principal chemistry used in making a polyurethane foam is the formation of urethane linkages from hydroxyl groups and isocyanate groups and the simultaneous generation of a gas that will enormously expand the reacting mass. This gas highly dilutes the polymeric network as it forms and remains as either open or closed gas cells in the final foamed product.

As the reaction proceeds, the gas cells grow both by diffusion of the gas molecules into the growing cells and by thermal expansion of the gas within the cells through the heat generated in the polymerization reaction. The foaming process is almost adiabatic. As it proceeds, the gas cells grow as expanding spheres; at some point in the process, these cells begin to contact one another. The gas cells, spherical at first, begin to pack in a close-packing array. As the cells begin to touch, their surfaces flatten and become thin as liquid drains from the contacting and thinning surfaces into the larger remaining space of the interstitial regions. The choice of a surfactant, almost always a silicone-based surfactant, is critical to the stability of the foam during processing and largely determines whether the foam cells remain closed or open at the time the foam reaches full expansion or its full rise-height (37).

At full rise, the liquid phase gels and thus produces the stable foam in its final, supramolecular architecture. If the system gels without the gas cell walls, or membranes, breaking, a closed-cell foam results. If the thinned membranes between the gas cells break at full foam rise and if the foam has developed sufficient mechanical integrity and stability, there is a final exhalation of gas from the foam and an open-cell foam results. (If the gas cells break without the foam having mechanical integrity, foam collapse occurs and an essentially solid polymeric mass results.) The success of the foaming process depends on the stability of the risen foam and whether the foam produced meets some very closely defined engineering properties. Meeting the required properties depends on very close control over the chemistry involved. The process of expansion of the components of the formulation during polymerization produces foam and reduces the density of the components from that of a urethane elastomer with a density of about 55 lb/ft^3 to that of a foam with a density of 1.5 to 3.0 lb/ft^3. This represents an expansion in volume of about 20 to 40 times.

The basic chemistry is the reaction of polyol and diisocyanate to give a crosslinked polyurethane (see Sections IV.A and IV.B).

$$R-OH \ + \ R'NCO \ \longrightarrow \ R-O-\overset{\overset{\displaystyle O}{\|}}{C}-\overset{\overset{\displaystyle H}{|}}{N}-R'$$

The polyols that are principally used in making polyurethane foams are poly(alkylene oxide)s, poly(propylene oxide), or poly(propylene oxide) containing about 5 to 25 percent of ethylene oxide units either as random or block copolymer (see Chapter 4, Sections IV and V). The ethylene oxide may be present as capped units, producing terminal primary hydroxyl units [poly(propylene oxide) polyol has almost all secondary terminal hydroxyls].

These polyols have functionalities of three or more terminal hydroxyls per molecule. The diisocyanates are most often TDI or one of the liquid forms of MDI (22). Formation of urethane linkages is catalyzed by organometallic catalysts that most often are tin compounds (23).

1. Rigid Foams

Rigid polyurethane foams were first used as the core of sandwich structures in aircraft. Polyester polyols were used in combination with TDI, employing the prepolymer technique described for elastomers in Sections IV.A and IV.B. Major developments in manufacturing rigid polyurethane foams came with the introduction of MDI as the isocyanate component and the use of chlorofluorocarbon blowing agents, especially fluorotrichloromethane.

The use of MDI, rather than TDI, especially MDI-prepolymer, simplified foam processing, and the use of fluorocarbon blowing agents with closed-cell foams resulted in low-density foams that had reasonably good mechanical properties and very good (low) thermal conductivities. The use of fluorocarbon blowing agents helped control the rapid temperature rise during foaming and greatly reduced the thermal conductivity of the foams in comparison with foams expanded by other gases such as air or carbon dioxide.

Specific conductivities are compared by K-factor, or simply K or k. The thermal conductivity is the time-rate heat transfer by conduction across unit area and through unit thickness for a unit difference in temperature. It is usually measured in $(cal)(cm)/(cm^2)(sec)(^{\circ}C)$. The heat, Q, conducted across a solid object of cross-sectional area A and thickness D in time t, when the opposite ends are at temperatures T_1 and T_2, can be expressed as

$$Q = KAt(T_2 - T_1)/D$$

where K is a constant termed the specific heat conductivity that is dependent on the nature of the substance being investigated. As mentioned above, K is often called the K-factor.

The poly(alkylene oxide)s are principally used to make rigid foams because they have good hydrolytic stability, are compatible with fluorocarbon blowing agents, and can be obtained with high functionalities, around 3 to 6, and with predetermined hydroxyl numbers (38). The polyols can be obtained by propoxylation of a starter (see Chapter 4, Section IV) such as glycerine (functionality of 3) or sorbitol (functionality of 6) or a number of amines such as ethylene diamine or phenylene diamine (functionalities of 4), with hydroxyl numbers typically in the range of 300 to 500 mg KOH/g (21).

Rigid foams are formulated with polyol, isocyanate, surfactant, catalyst, blowing agent and sometimes filler, an additional crosslinking agent (chain-extender) as described in Sections IV.A and IV.B, and flame retardant. Isocyanate indexes are usually high—in the range of > 150 and up to 300. Isocyanate index is 100 times the ratio of equivalents of isocyanate to the equivalents of reactive hydrogen compounds (e.g., hydroxyl or amine compounds). If there is a stoichiometric equivalence of isocyanate to reactive hydrogen to isocyanate, the ratio is unity and the index is 100. If the index is 200 , there are twice the number of isocyanate groups as there are reactive hydrogen groups.

At very high indexes, the foam made is a combination of isocyanurate and urethane structures. Isocyanurate is the trimer of isocyanate.

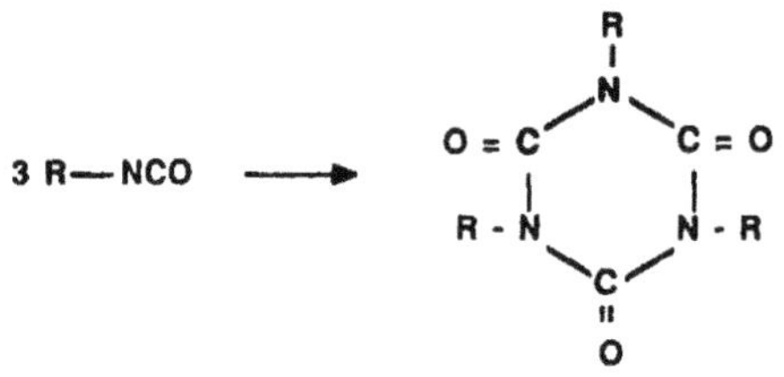

The isocyanurate ring has high thermal stability and high combustibility resistance. The isocyanurate, in reality, is trifunctional and also contributes to foam-matrix crosslinking, be-

cause the isocyanates used are di- or higher functionality isocyanates. Thus, each R group in the structural formula above
contains an isocyanate or part of another isocyanurate ring.
Urethane-isocyanurate rigid foams based on low-molecular-weight
propylene oxide polyols are the principal commercially used rigid
urethane foams.

Rigid foams can be processed as continuous panels or by
lamination between two surfacing materials (sandwich panels
with a rigid polyurethane foam core). Composite sandwich panels may have particleboard or plasterboard on one surface and
cement on the other. Foams can also be produced to fill cavi-
ties between metal sheets or sheets of reinforced plastics or in
combination on either side. These structures are used as
thermal insulation for refrigerator and freezer cabinets, refrig-
erated transportation vehicles, water heaters, pipes, pipe shells,
tanks for gas storage or transport of liquified natural gas, and
so on.

2.   Flexible Foams

Flexible polyurethane foams are principally characterized by
physical properties of density, compression set or indentation
modulus, and resiliency. These properties are important in end
uses such as cushioning in home, office, and automotive seating; fabric backing and insulation; carpet underlay; and packaging. The end-use requirements of these applications relate
well to the elastomeric and load-bearing properties of such
foams.

Basically, the chemistry of making flexible foams is similar
to that used to prepare polyurethane elastomers and rigid foams.
However, the chemistry has some added features and requirements for control over the reaction processes that are particularly severe (20).

Basically, flexible polyurethane foams are formed from
polyols and isocyanates. The polyols have hydroxyl functionalities of at least 2 if the foam is to be made by a prepolymer
process and of 3 or more if the foam is to be made by a "one-
shot" process. The prepolymer process proceeds in two distinct steps; while in the one-shot process, all reactants are
intensively mixed together at the beginning of the foaming
process. Early in the development of urethane foams, the polyols were polyester diols; now, however, by far the largest volume of polyols for urethane foams are poly(propylene oxide)
polyols or poly(propylene oxide) polyols containing up to 25

percent ethylene oxide either as a random comonomer, as a block comonomer, or as an oxyethylene cap to provide terminal primary hydroxyl groups.

In the case of the prepolymer process, polyols with molecular weights in the range of 1000 to 2000 are usually used. In the one-shot processes, polyols have molecular weights of at least 3000 and may have molecular weights in the range of 5000 to 12,000 or more, with average polyol hydroxyl functionalities of 3 to 6. Blends of different molecular weight polyols are used to design foams with optimum properties for certain end uses.

As in the case of urethane elastomers and microcellular polymers, the isocyanates used are generally TDI, MDI, and the liquid MDIs (22), which have been described above. Blends of these isocyanates have also been used.

When formulating a flexible foam, the hydroxyl number of the polyol or polyol blend and the equivalents of isocyanate used must be known with high accuracy. The ratio of the reactants must be accurately metered when mixing the formulation components, just before pouring the chemicals for the foam formation.

The basic chemistry of the reactions taking place during foaming are formation of urethane linkages from hydroxyl and isocyanate groups

$$R-OH \quad + \quad R'-NCO \quad \longrightarrow \quad R-O-\overset{\overset{\displaystyle O}{\|}}{C}-\overset{\overset{\displaystyle H}{|}}{N}-R'$$

polyol   isocyanate   urethane

the sequential reaction of isocyanate with water to form an amine and carbon dioxide by breakdown of a transient species, carbamic acid,

$$R-NCO \quad + \quad H_2O \quad \rightarrow \quad [R-NH-COOH] \quad \rightarrow \quad R-NH_2 \quad + \quad CO_2$$

isocyanate   water   carbamic   amine   carbon
                                acid   dioxide

and the reaction of the amine formed with another isocyanate to form a urea linkage.

$$R{-}NH_2 \quad + \quad R'{-}NCO \quad \longrightarrow \quad R\overset{H}{\underset{|}{-}}N\overset{O}{\underset{\|}{-}}C\overset{H}{\underset{|}{-}}N{-}R'$$

amine   isocyanate   urea

The process of converting the mixture of polyols and isocyanates to foams takes only a few minutes. The reactions are
essentially adiabatic. The foams produced must meet the close
tolerances of highly engineered materials. Very close control
over the sequence and timing of these reactions during foaming
is achieved by catalyst selection, polyol reactivities, and concentration of reactants, surfactants, and other proprietary ingredients in the foam formulation.

In the early 1950s, flexible foams were made by the prepolymer process. In this process, a prepolymer is prepared by
reaction of a diisocyanate with a diol, usually a linear polyether
diol similar to those described in Section IV.A for elastomers.
The prepolymer polyether diisocyanate, usually a poly(propylene
oxide) diisocyanate, has a molecular weight in the range of 1000
to 2000. Water is used as a chain extender in making the prepolymer and also to provide branch sites to increase the functionality of the prepolymer through formation of biuret linkages.

$$HO{-}R{-}OH \quad + \quad OCN{-}R'{-}NCO \quad + \quad H_2O \quad \rightarrow$$

polyether   diisocyanate   water
polyol

$$OCN{-}R'{-}\overset{H}{\underset{|}{N}}{-}\overset{O}{\underset{\|}{C}}{-}\overset{H}{\underset{|}{N}}{-}R'{-}\overset{H}{\underset{|}{N}}{-}\overset{O}{\underset{\|}{C}}{-}O{-}R{-}O{-}\overset{O}{\underset{\|}{C}}{-}\overset{H}{\underset{|}{N}}{-}R'{-}NCO \; + \; CO_2$$

isocyanate-terminated, urethane-urea, extended prepolymer

extended prepolymer   +   diisocyanate $\rightarrow$

$$OCN{-}R'{-}\overset{H}{\underset{|}{N}}{-}\overset{O}{\underset{\|}{C}}{-}N{-}R'{-}N{-}\overset{O}{\underset{\|}{C}}{-}O{-}R{-}O{-}\overset{O}{\underset{\|}{C}}{-}\overset{H}{\underset{|}{N}}{-}R'{-}NCO$$

C=O

N−H

R'−NCO

isocyanate-terminated, biuret prepolymer

This product can then react with water, a reaction that proceeds at a higher rate with tertiary amine catalysts, to produce a foam blown by carbon dioxide and crosslinked mainly through biuret linkages. Further chain extension with branching and crosslinking occurs during foaming, with formation of more biuret and allophanate linkages (Chapter 5) formed by reaction of isocyanate groups with urethane linkages.

The one-shot process for making urethane foams is so named because all reactants are mixed at once rather than in two steps with intermediate isolation of prepolymer. On a molecular scale, the one-shot process is distinguished from the prepolymer process in that crosslinking is principally through the urethane and urea linkages, rather than through the less hydrolytically and thermally stable biuret and allophonate linkages. The simplicity of the one-shot process has led to its largely supplanting the prepolymer process for making urethane foams. However, the one-shot process has much closer requirements for process control.

Control of the reaction sequence in making flexible, open-cell, polyurethane foam using the one-shot process is accomplished by choosing the polyol, the surfactant, and the combination of tertiary-amine and organometallic, usually tin, catalysts. As in the previously described procedures for making urethane elastomers and microcellular products, for foams, polyols—poly(propylene oxide) polyols often containing minor amounts of ethylene oxide—are used. These polyols are selected by functionality, molecular weight, and reactivity (see Table 3).

Silicone surfactants are the class of surfactant materials used almost exclusively in making urethane foams. A large amount of technology has been accumulated relating to the design and selection of particular silicones for the two major methods of producing water-blown foams, which are molded foams and slab foams. As the term implies, molded foams are produced by intensive mixing of the formulation ingredients, which are then poured into molds. The foam rises in the closed mold and, on removal, has essentially its final end-use shape. Foams made in this way are used extensively for automobile seats, headrests, crash pads, and so on. Slab foams are poured from the intensive mixer into large paper troughs that move on conveyors away from the pouring site as the foam rises. These foams, which are called buns, may be 100 or more feet long, 5 or 6 feet wide, and 5 or 6 feet high after foam rise (Figure 2).

FIG. 2 Slab-stock foam that is in the process of foaming. Bun
has exited pouring area and is moving toward the viewer.

Generally, stronger surfactants are used in making slab foams
than are used in molded-foam formulations.

The following description is clearly an oversimplification,
and more detailed descriptions can be found in the literature
(20, 23, 36, 39). The rections and reaction sequence taking
place during the process of making water-blown polyurethane
foam can be described as follows.

The tertiary-amine catalysts principally affect the reaction
of water with isocyanate

$$R-NCO + H_2O \xrightarrow{\text{t-amine}} [R-NH-C-OOH] \rightarrow R-NH_2 + CO_2$$

isocyanate   water   carbamic acid   amine   carbon
                                                                  dioxide

while the tin catalyst principally increases the reaction of hydroxyl with isocyanate.

$$R-OH \quad + \quad R'-NCO \quad \xrightarrow{Sn} \quad R-O-\overset{\overset{\displaystyle O}{\|}}{C}-\overset{\overset{\displaystyle H}{|}}{N}-R'$$

polyol   isocyanate   urethane

The concentration and the relative amounts of the two catalysts control the rate of rise of the foam (carbon dioxide generation) and the rate at which the foam elastomer network, or gel, forms.

In a simple foam formulation for a water-blown foam of polyol, diisocyanate, water, silicone surfactant, and tertiary-amine and stannous octoate catalyst, the first reaction to occur is water with isocyanate to liberate carbon dioxide, which nucleates to form gas cells and begins foam expansion. This reaction of water with isocyanate is the most energetic reaction that will occur, and it has about twice the heat of rection of either amine with isocyanate or hydroxyl with isocyanate.

The primary amine (with either TDI or MDI) reacts with more isocyanate to form urea linkages that will coalesce or aggregate to form polyurea domains in the rising foam that, at this stage, is still largely liquid-phase reactants. Near full rise of the foam, urethane formation becomes the dominant reaction to produce the elastomeric network of the foam. The solid phase of the fully cured foam can be described as a segmented block urethane-urea copolymer. The continuous phase of the elastomeric foam is polyether urethane reinforced by polyurea domains that form the discontinuous, high-softening, reinforcing phase of the elastomer network. At the full rise-height of the foam, the walls of the gas cells, which have been expanding due to heat of reaction, break open in the case of an open-cell, flexible foam. The cell-opening process is complicated and depends not only on the surfactant used and its concentration but also on the process of aggregation of the polyureas as these domains form. The actual supramolecular architecture of the foam is determined first by the polyureas, which form a gel of hydrogen-bonded structures. The urethane reaction chemically gels the elastomer network in the final form.

The major applications for flexible slab-stock polyurethane foams are furniture cushioning, both seating and backs; seating in public transportation; textile backing; carpet backing for tufted carpeting; and home carpet underlay. A considerable amount of flexible foam is used in packaging, particularly in semipermanent containment of instruments and tools.

Molded foams are used chiefly in automotive seating and composite structures with metal, wood, or plastic inserts. Molded foams are also used for sound barriers and vibration insulation in automobile interiors and carpets.

In 1985, world production of flexible polyurethane foams was about 3 billion pounds (1.4 million metric tons), with about half of that production and consumption in the United States. Of that total volume, approximately two-thirds, or 2 billion pounds (900,000 metric tons), was poly(alkylene oxide).

V. OTHER APPLICATIONS

Although the above discussion details the chemistry of poly(alkylene oxide)s when used in the major applications of surfactants, polyurethane elastomers, RIM parts, and foams, there are a number of other end uses. Space constraints do not allow an in-depth discussion of these end uses, but the following examples are offered of the broad and interesting uses to which these polymers are put.

A. Adhesives

Poly(ethylene glycol) is used in adhesives to prevent hardening, as a humectant, and to provide lubricity. The polyglycols are used with starch, dextran, casein-based adhesives, and collagen-based glues (animal glues), to prevent drying and cracking. In poly(vinyl acetate) latex adhesives, they are used as plasticizers. Small amounts of a 4000-molecular-weight polyethylene glycol are added to an adhesive for postage stamps to reduce sheet curl (40).

A water-dispersible, hot-melt adhesive for recyclable paper and paper products such as books can be made by mixing a small amount of a tackifier with a vinyl acetate grafted poly(alkylene oxide) (41). The grafted polyether is 40 to 80 percent by weight vinyl acetate grafted onto a poly(alkylene oxide) that has a molecular weight of 3000 to 20,000 and that is at least 50 percent oxyethylene. The adhesive is water dispersible, a key advantage when the paper product is to be reprocessed to produce a good-quality paper. The grafting reaction takes place during free-radical-initiated (benzoyl peroxide), bulk polymerization of vinyl acetate with the preformed poly(alkylene oxide) dissolved in the vinyl acetate monomer. The product has a viscosity in the range of 15,000 cP at 350°C.

With paper products that are wound on a core, such as
paper towels and toilet tissue, an adhesive is needed to adhere
the paper to the roll-core at the start of the roll and hold the
last part of the roll in place at the end (called pick-up and
tie-end adhesives). Water solutions of high-molecular-weight
poly(ethylene oxide), combined with some filler and with or
without a colorant, are effective because the pituitousness of
these solutions provides the needed tack with sufficient wet
strength for packaging purposes without transferring through
several layers of the paper product. When these solutions dry,
they lose most of their adhesive properties, and tearing off excess paper from either end of the roll is largely eliminated.
The very dilute concentration of high-molecular-weight poly-
(ethylene oxide) used leaves the paper in its normal, soft condition at both ends and thus eliminates the hard glue spots that
other adhesives might leave, making the end sections of the
roll unusable. Exact formulations are proprietary.

A number of adhesive compositions have been developed
based on molecular-association complexes of high-molecular-
weight polymers of ethylene oxide. These include (1) water-
borne wood glues, which are based on poly(ethylene oxide)
with molecular weights in the range of 500,000; (2) catechol
lignins, water-soluble, quick-set adhesives based on novolac-
phenolic resins and poly(ethylene oxide); and (3) baked-on,
solvent-borne, pressure-sensitive adhesives made from novolac-
phenolic complexes of poly(ethylene oxide) (see Table 2).

B. Cosmetics and Toiletries

Polyethylene glycols are formulated into creams, lotions, jellies,
and cakes to impart a skin-smoothing feel and moisturizing qual-
ities. They are formulated into vanishing creams, brushless
shaving creams, toothpastes, suntan lotions, hair shampoos and
dressings, solid perfumes, and solid antiperspirants.

In shaving creams, a few tenths of a percent of a high-
molecular-weight poly(ethylene oxide) will prevent the shaving
cream from drying out or becoming stiff if shaving is interrupt-
ed for a few minutes. The presence of low levels of poly(ethylene oxide) provides a lubricity that makes shaving easier and
makes the skin feel smoother after shaving. The same level of
poly(ethylene oxide) in an after-shave lotion provides a very
pleasant, smooth, fresh feel to the skin. In a like manner,
very low concentrations of poly(ethylene oxide) in a rubbing

FIG. 3 Strip (white area) of poly(ethylene oxide) visible on a
commercial, disposable razor.

alcohol compound, isopropanol- or ethanol-based, enhances the
smooth feel of a massage and gives the skin, after the massage,
a full, smooth, soft feel (42). One manufacturer has added a
strip of solid poly(ethylene oxide) into the shaving head of one
of its disposable razor blades (Figure 3). The polymer slowly
dispenses, by leaching, a small amount of polymer onto the
face during shaving. The polymer provides a lubricating action
and a "microsoft" feature (43).

C. Compounding, Processing, and Formulation Materials

Poly(ethylene glycols) are used as fiber-processing lubricants
(44). High-molecular-weight poly(ethylene oxide) is used as a
sizing for glass fibers, and intermediate-molecular-weight
grades are used as a warp size for cotton textiles (45). High-
molecular-weight poly(ethylene oxide) is used as an antistatic
additive during melt-spinning of polyester and polyamide fibers.
Copolymers of alkylene oxides and lactones have been used as

dye assistants, antistatic agents, and other property enhancers
in polyolefin, cellulosic, and nylon fibers (46—48).

High-molecular-weight poly(ethylene oxide) monofilament has
been used as fugitive textile weft. At predetermined intervals
during weaving, the fugitive weft is introduced. When the
weaving process is completed and the textile is washed, the
water-soluble, fugitive weft is removed, and the spaces left by
its removal define and facilitate neater cutting or tearing of the
fabric at the predetermined locations.

In the processing of rubber and plastic, polyethylene glycols are used for heat-transfer baths in rubber vulcanization
and plastics forming, where temperatures of 150°C to 350°C are
required. These glycols are used as release agents in molded
rubber products. In compounding hydrocarbon rubber, polyethylene glycols are used as carbon black dispersants. In sili-
cate-reinforced rubber, polyethylene glycols are used as compounding lubricants and to coat the silica to prevent the curing
accelerators from being absorbed by the silica and becoming
ineffective in the compounding formulation.

In paper products, polyethylene glycols are added to clay
and starch to promote paper smoothness and gloss. These
glycols are used as humectants and softeners for leathers.

Polyethylene glycols are formulated into ceramics to improve
"green strength" and as binders for the ceramic. These materials have the important advantage of completely or cleanly
burning and leaving no char or discoloration in the final product. Poly(ethylene oxide) with a molecular weight in the range
of 1 million has the advantage of providing superior binding
power and also superior lubricity for the ceramic particles during forming.

In metal working, polyethylene glycols are used as lubricants; as rolling, cutting, and grinding fluids; in soldering
fluxes; and during buffing and polishing, because of desirable
lubricity, low change of viscosity with temperature, and low
corrosion of metal surfaces. The higher glycols and waxes are
used in lost-wax metal casting.

High-molecular-weight poly(ethylene oxide) is used to thick-
en solutions that are used to clean surfaces. The thickened
solutions have a syrupy or rubbery texture that holds the solution on the surface to be cleaned and prevents runoff or
drip-off. Examples are cleaning vertical surfaces with muriatic
acid, sanitary cleansers for vitreous surfaces such as toilet
facilities, etc. High-molecular-weight poly(ethylene oxide) is

used with citric acid, oxalic acid, phosphoric acid, hydrochloric acid, and ammonium hydroxide. The technique cannot be used with sodium or potassium hydroxide because the poly(ethylene oxide) is not soluble in water at high pH (see Chapter 6) (49).

A high-molecular-weight poly(ethylene oxide) is used at concentrations of 2 to 60 lb/10,000 gal. to modify bituminous emulsion in paving compositions. The high-molecular-weight poly(ethylene oxide) controls the flow of the emulsion around aggregate and thus aids in binding the aggregate and improving paving uniformity and performance (50).

Water-soluble films are made from poly(ethylene oxide) with a molecular weight of about 500,000 by melt-processing manufacture that employs either a calendering or a blown-film operation. A process for producing biaxially oriented film from poly(ethylene oxide) has been described (50a). The films, usually 1 to 3 mils thick, have been used as seed tapes for planting crops including lettuce, onions, and celery and as packaging materials for insecticides, fertilizers, and dyestuffs. High-molecular-weight poly(ethylene oxide) has been used in the manufacture of battery electrode separators (51).

Polyethylene glycols are used in electroplating to improve smoothness and grain uniformity. High-molecular-weight poly(ethylene oxide) has been used as a temporary binder for phosphors on the glass tube in the manufacture of fluorescent lamps (52).

D. Water Treatment and Flocculation

An important use of very-high-molecular-weight poly(ethylene oxide), molecular weights of several million and more, has been as flocculating agents. This use has been increasingly exploited with the increasing emphasis that is placed on cleaning up waste streams from industrial processing and on municipal water treatment. An important feature of the use of poly(ethylene oxide)s as flocculants is their very low biological oxygen demand (BOD).

An example of the use of poly(ethylene oxide) flocculants and coagulants is their use in aqueous clay suspensions (53). Also, coal is washed after mining to remove coal dusts. Treatment of the water effluent, which is dark colored and often acidic, with poly(ethylene oxide) is an effective method for removal of the fine particles (54). Other mining operations produce silica-contaminated waste streams that require treatment

(55). Poly(ethylene oxide) is used at various stages in the paper-making process. It is useful for retention of the fillers and sizings that are added in the paper-making process and later to clean up white-water paper-mill effluent (56).

E. Hydrodynamic Drag-Reduction Agents

One of the most intriguing properties of high-molecular-weight poly(ethylene oxide) in water solution is the ability of very small quantities of the polymer to affect enormously the friction or drag reduction of water when it flows or is pumped through a pipe or hose or over a surface. This drag reduction is called the Tom's Effect and is described in Chapter 6.

This effect of reducing the friction of water flowing across a surface has been demonstrated on boat hulls. It is against racing rules to use such friction-reduction agents in boating competition. Analytical techniques, including one using the poly(acrylic acid) molecular-association complex, have become standardized for use in enforcing this rule in international sailing competitions. The effect has been used to increase the speed of naval craft and torpedoes.

In fire-fighting, poly(ethylene oxide) has been used to reduce the size of firehoses while maintaining the water delivery of a larger hose. This gives fire fighters greater mobility and flexibility, which is especially important in fighting fires in high-rise buildings where the heavy and clumsy water hose has to be manipulated up stairwells from floors below the fire. Introducing small quantities of the polymer into the water stream also permits the pumping of a greater volume of water, or increasing the "throw" of water, from a fire truck hose pumper (58). In the latter case, a proprietary device can be used to introduce the readily dispersible/soluble polymer into the water stream at the fire truck.

Water-jets with nozzle pressures of 30,000 to 60,000 psi (200-400 MPa) are used in the fluid-jet cutting of textiles, rubber products, foams, and cardboard. When small amounts of high-molecular-weight poly(ethylene oxide) are introduced into the water stream before the nozzle, the jet maintains a cohesive stream and cutting is greatly improved (59).

Poly(ethylene oxide) additive can prevent sewer surcharges in storm sewers during heavy water flows, such as during a flash flood condition. The storm sewer system may be inadequate to handle such overloads. The alternative of building a larger volume system to replace an existing sewer system may not be attractive for a community facing only infrequent overloads. A proprietary system has been developed to inject a formulated poly(ethylene oxide) solution into the system to increase volumetric flow and the existing system's capacity during emergencies (60).

F. Photography, Lithography, Printing Inks, and Coatings

In photography, poly(alkylene oxide)-based antifog agents are used for silver halide emulsions. Polyethylene glycols are formulated into photographic developers to improve contrast and speed. The sulfate ester of a butynediol-polyether is used to prevent loss of sensitivity during film storage (61). A block copolymer poly(alkylene oxide) is added to photographic emulsions to prevent development of graininess. The preferred block copolymer has oxyethylene units on either end and oxypropylene units in the center of the polymer chain. The end, ethylene oxide blocks have a degree of polymerization from 4 to 48, and the center oxypropylene units have degrees of polymerization from 14 to 52.

Polyethylene glycols are formulated into the thixotropic inks used in ballpoint pens. They are also used in stamp-pad inks to retain water and, thus, good flow characteristics over long periods of time. Polyethylene glycols are used in formulating steam-set inks. High-molecular-weight poly(ethylene oxide) in combination with poly(acrylic acid) forms a molecular-association complex that can be used to microencapsulate nonaqueous inks. An oil-in-water emulsion of the ink is first made at a high pH with poly(ethylene oxide) and poly(acrylic acid) present. The pH is then reduced to the point at which the water-insoluble complex forms at droplet interfaces, with microencapsulated ink resulting (see Chapter 6, Figure 6.3).

Solutions of high-molecular-weight poly(ethylene oxide) in water are too stringy for the polymer to be used as a thickener for latex paints. However, the high polymer can be used to achieve special decorative effects such as spatter finishes and decorative pattern effects in spray finishes (62). Polyurethanes are used as architectural cabinet and furniture finishes (63).

Polyethylene glycols are used to modify alkyds and polyesters to obtain water-dispersible coatings. High-molecular-weight poly(ethylene oxide) is used to thicken chlorinated solvents that are used for paint and varnish removal (64).

G. Wood Products and Artifact Preservation

Selected woods are often used for their decorative effects.
These effects include burls, high coloration, and attractive
grains such as those found in wild cherry wood and are often
prized by woodworkers for such decorative items such as clock
faces.   To prevent warping or splitting of finely worked pieces,
the wood can be soaked in poly(alkylene oxide)s prior to working.

Liquid polyethylene glycols and polypropylene glycols are
used as swelling agents to tighten wood joints in cabinetwork.
PEG-1000, a 1000-molecular-weight polyethylene glycol, is particularly useful as a wood impregnant to prevent shrinking,
cracking, and warping after drying.   It notably improves
workability of the wood after drying.   Products such as PEG-
1000 are available from commercial woodworking supply houses.

A use of concentrated, aqueous solutions of polyethylene
glycols is in naval archeology.   Examples in the last few years
have been the raising and preservation of ships from Chesa-
peake Bay, the Mary Rose in England (65), and the Wasa in
Stockholm Harbor (66, 67).

In the simplest terms, wood is a composite structure consisting of strengthening and toughening materials—the cellulosic
polymers, cellulose and hemicellulose—and a brittle, crosslinked,
cementing material—lignin.   Lignin is highly complex and difficult to classify accurately, though it contains many aromatic
residues.   When buried shallowly in normal soil under aerobic
conditions, wood is deteriorated by a combination of chemical
and biological attack.   However, wood contained in ships that
have sunk in deep water or are held in other essentially oxy-
gen-free environments deteriorate very, very slowly through
attack of anaerobic bacterial that feed on the cellulosics.   After
very long periods of time under such conditions, only a skeleton of lignin will remain.

The wood structure after long immersion—which, in the
case of archaeological wood, may be hundreds or thousands of
years—is swollen.   The lignin and ash content of the wood increases relative to the cellulosic portions as immersion time increases.   The skeleton of waterlogged, swollen lignin and any
remaining cellulosics has free water within the cells.   This
water is strengthening and prevents collapse.

The wood structure has a shape that is very reminiscent of
the original shape.   However, after recovery and during a subsequent drying operation that takes place without treatment, the

free water is lost and the originally wooden archaeological objects warp, splinter, and lose shape and integrity because the remaining skeleton cannot withstand the stresses and accompanying strains that are involved in the drying process.

In the case of the ship _Wasa_, which was 230 feet long with a mainmast of 170 feet and a displacement of 1300 tons, the ship heeled in a swift gust of wind on her maiden voyage from Stockholm Harbor on Sunday afternoon, August 10, 1628, and sank in 110 feet of water. Three hundred thirty-three years later, the ship was raised and a painstaking restoration was begun. The process, involving the waterlogged hull, 15,000 pieces of disconnected wood, and 700 wood sculptures, took 17 years.

The _Wasa_ was well preserved for a ship submerged for 300 years. However, the waterlogged wood had changed composition and lost its ability to maintain its dimensions and any of its strength if dried. Scientists charged with the preservation and restoration used various polyethylene glycols during the treatment process. All small wooden pieces were treated for 18 months in tanks of aqueous polyethylene glycol containing 2 percent boric acid. A dilute aqueous solution (10 percent) of a 4000-molecular-weight polyethylene glycol was used principally; over time, the polyethylene glycol concentration was increased to 30 percent.

When the aged, wet, and soaked wood first contacts the aqueous solution of polyethylene glycol, water in the wood is gradually replaced with the polyethylene glycol molecules by diffusion. The diffusion process depends on concentration, molecular weight or size and shape of the glycol molecules, temperature, and agitation, as well as on porosity and tortuosity of the structure being treated. Porosity depends on the wood species as well as the wood type—hard or soft wood—and tortuosity depends on both wood species and extent of deterioration (68, 69). After a time, the glycol concentration in the treatment solution is increased to maintain a reasonable rate for the diffusion process. Gradually, the voids are filled with the polyethylene glycol and the structure is strengthened without depending on water for support. For items that are too large to immerse, such as the oak timbers of the _Wasa's_ main hull, spray techniques with a relatively low-molecular-weight (600) glycol are used.

The polyethylene glycol molecular weight was gradually increased as the treatment solution's concentration was gradually

increased from 10 to 45 percent. The complete drying and restoration process for the Wasa's hull is expected to take decades.
It is believed that surface layers of absorbed polyethylene glycol, as well as diffused glycol from the sprayed material, will
regulate the drying process and prevent cracking, warping,
and structural collapse.

The polyethylene glycol treatment to preserve the Wasa has
been followed in England in preservation of the British warship
Mary Rose, the intended flagship of the navy of Henry VIII,
which also sank on its first voyage from harbor. Polyglycol
wood preservation has been applied to a number of ship relics
of the U.S. Civil War period that have been raised in Chesa-
peake Bay and along the Carolina coasts and represents an interesting application of the compatibility and molecular complexing properties of the poly(alkylene oxide)s.

Other examples of the use of polyethylene glycols to preserve artifacts include leather restoration (70—72) and consoli-
dation of the surface regions of limestone sculptures (73).
These applications use neat, solid glycols, with different molecular weights used in different areas. Wood artifact preservation also includes paleolithic artifacts from Zimbabwe (74), which
were some of the first to be treated by the aqueous polyethylene glycol process first introduced by Stamm in 1956 (75). At
the Cleveland Museum of Natural History, a 3500-year-old dug-
out canoe that was found in a northern Ohio peat bog is still
undergoing treatment (76). It is interesting to point out that
other water-soluble polymers have been investigated for the
preservation of wood, but the results were less satisfactory
than those obtained with polyethylene glycols (66).

H. General and Specialty Areas

Thermoplastic elastomers, of which polyurethanes, polyureas,
and polyurethane-ureas are important factors, are rapidly replacing thermoset rubbers and metals, as well as entire assem-
blies made of these and other materials of construction. The
thermoplastic elastomer market is greater than 1 billion pound
per year and is growing at a rate of 9 percent per year (77).
The broad utility of thermoplastic elastomers in the creative design of products for a number of end uses, such as automotive
parts, athletic footwear, seals, jewelry, etc., has been described by Sheridan (78).

A specific compound, Polyethylene glycol 1450, with a molecular weight of 1300—1600, has been found to be a superior

fusogen for lymphocyte-myeloma hybrids (79). The glycol plus
dimethylsulfoxide and phosphate-buffered saline solution are
mixed to form the fusogen.

Relatively high-molecular-weight polyethylene glycol,
CARBOWAX^{TM} 20M, is widely used in gas chromatography columns. This compound allows separation and resolution of polar
compounds that have similar boiling points. The glycol had
several deficiencies—such as low thermal stability, 220°C, poor
phase stability; and a high minimum operating temperature,
60°C—that have been a source of difficulty for many years.
Currently, the deficiencies have been overcome by bonding this
glycol to the capillary columns. With the bonded-glycol columns,
operating temperatures as low as 20°C can be used, thermal
stability has been increased to 280°C, and oxidative suscepti-
bility has been overcome (80). In the area of gel permeation
chromatography, poly(ethylene oxide) is used as a standard and
is available in molecular weights of 18,000 to 996,000 (81).

Polyepichlorohydrin and poly(propylene oxide) rubbers have
been discussed recently (82). World usage of polyepichlorohydrin rubber is 20—22 million pounds (9—10,000 metric tons) per
year and has a growth rate of 7—8 percent per year. It is
used mainly in the automotive industry, where advantage is tak-
en of polyepichlorohydrin's excellent ability to withstand ozone
and heat and its good air- and oil-permeability characteristics.
Worldwide demand for poly(propylene oxide) rubber is about 1—
2 million pounds (450—900 metric tons) per year. This rubber
is effective in high-performance tubing applications that need
both high-temperature resistance and the properties of natural
rubber. Poly(propylene oxide) rubbers have upper use-tem-
perature limits of 145°C, compared to 110°C for natural rubber.

Recently, phase-change clothing has been described (83).
This clothing is made from a superior insulating fabric that is
composed of a network of polymers that can change characteristics (phases) as ambient temperatures change. To achieve
the effect, which lasts for about 30 minutes, the network of
polymers is impregnated with an aqueous solution of a 1000-
molecular-weight poly(ethylene glycol), PEG 1000, and dried.
The PEG is the phase-transformation compound. When the temperature increases, the PEG melts and takes on heat, giving a
cooling effect to the body. At the same time, the fabric sof-
tens. When the temperature falls, the PEG crystallizes and re-
leases heat to warm the body. At the same time, the fabric
stiffens. The fabric is designed for the sportswear market,
and tests with volunteers wearing skiwear found the clothing

to be comfortable, launderable, durable, and capable of providing warmth and cooling while absorbing body moisture.

REFERENCES

1. W. D. Willis L. O. Amberg, A. E. Robinson, and E. J. Vandenberg, Rubber World 153:88 (1965); D. A. Butler and E. J. Vandenberg, in Handbook of Elastomer: New Developments and Technology (A. K. Bhomick and H. L. Stephens, eds.), Marcel Dekker, New York, 1988, p. 163.
2. A. E. Winslow and K. L. Smith, U.S. Patent No. 3,125,544 (1964); Forming Association Compounds, Union Carbide Corporation, Technical Brochure F-43272.
3. A. Lantella, Modern Paint and Ctgs, p. 28 (Feb. 1989).
4. Surfactant Science Series, Vols. 1 to 26 and later volumes, Marcel Dekker, New York, 1966–1987.
5. M. Daeuble, K. Oppenlaender, and R. Flickentscher, U.S. Patent No. 3,830,677 (1974).
6. G. Schroder and E. Brehmer, U.S. Patent No. 4,102,845 (1978).
7. W. Pilcher and S. L. Eaton, U.S. Patent No. 2,999,068 (1961).
8. B. Burczyk, A. Piasecki, and A. Sokolowski, Pol. PL 131,509 (1986); C.A. 110: 58337 (1989).
9. J. Halko, U.S. Patent No. 3,699,057 (1972).
10. C. Andrei, L. Bardasu, T. Kassovitz, and R. Messinger, Rom. RO 94,051 (1988); C.A. 110: 58366 (1989).
11. Res. Dev. Corp. Japan, EP Application 217,062-A (1985).
11a. R. K. Drummond, J. Klier, J. A. Alameda, and N. A. Peppas, Macromolecules 22:3818 (1989).
12. W. G. Swafford, Am. J. Hosp. Pharm. 19:134 (1962).
13. P. A. King, U.S. Patent No. 3,783,872 (1974).
14. American Pharmaceutical Association, Handbook of Non-Prescription Drugs, 1971.
15. H. B. Lee and D. T. Turner, J. Biomed. Material Res. 11:671 (1977).
16. "Camouflaged Drugs Make Enzon Stand Out," Business Week, April 27, 1987, p. 115.
17. J. G. Bots, L. van der Does, A. Bantjes, and B. Lutz, Br. Polymer J. 19:527 (1987).
18. V. Sa da Costa, D. Brier-Russel, G. Trudel III, F. D. Waugh, E. W. Salzman, and E. W. Merrill, J. Colloid/ Interface Sci. 76:596 (1980).

19. V. Sa da Costa, D. Brier-Russel, E. W. Salzman, and E. W. Merrill, J. Colloid/Interface Sci. 80:445 (1981).

20. F. E. Bailey, Jr., "Flexible Polyurethane Foams," to be published in Handbook of Polymeric Foams and Foams Technology (D. L. Klempner and K. C. Frisch, eds.), Hanser Publishers, New York.

21. Hydroxyl number is described by ASTM 2849-69 and, in Europe, by DIN 53240.

22. P. Fischer and E. Meisert, U.S. Patent No. 3,152,162 (1964); T. R. McClellan and R. A. Kolakowski, U.S. Patent No. 3,394,164 (1968).

23. J. H. Saunders and K. C. Frisch, Polyurethane Chemistry and Technology, Part I, Interscience Publishers, New York, 1967.

24. S. H. Metzger and D. J. Prepelka, in Advances in Urethane Science and Technology, Vol. 4 (K. C. Frisch and S. L. Reegen, eds.), Technomic Publishing Company, Westport, Connecticut, 1976, p. 141.

25. C. W. Macosko, Fundamentals of Reaction Injection Molding, Hanser Publishers, New York, 1989.

26. J. Harries, Kunstoffe 59:398 (1969); F. W. Pahl and K. Schulter, Kunstoffe 61:540 (1971); H. Wirtz, J. Cellular Plastics 2:324 (1966).

27. J. A. Faucher, Polymer Ltrs. 3:143 (1965).

28. P. J. Flory, Principles of Polymer Chemistry, Cornell University Press, Ithaca, 1953.

29. M. C. Pannone and C. W. Macosko, J. Appl. Polymer Sci. 34:2409 (1987).

30. D. Nissen and R. A. Markova, J. Elast. Plast. 15:71 (1983).

31. J. P. Casey, B. Milligan, and M. J. Fascolka, J. Elast. Plast. 17:218 (1985).

32. R. P. Taylor, J. E. Dewhurst, and A. M. Abouzahr, U.S. Patent No. 4,442,235 (1984).

33. R. J. G. Dominguez, J. Cellular Plast. 20:433 (1984).

34. R. J. G. Dominguez, D. M. Rice, and R. F. Lloyd, U.S. Patent No. 4,396,729 (1983).

35. P. Kolodziej, W. P. Young, C. W. Macosko, and S. T. Wellinghof, J. Polymer Sci., Phys. 24:2539 (1986); M. Bergmann, U. Maier, H. Muller, and L. Pierkes, Conference Handbook, 13th Inst. fur Kunstoffverarbeitung, Aachen, 1986, p. 405.

36. G. Oertel, ed., Polyurethane Handbook, Hanser Publishers, New York, 1985.

37. B. Kanner, B. Prokai, C. S. Eschbach, and G. J.
     Murphy, J. Cellular Plastics 15:315 (1979); B. Kanner,
     S. Drap, and G. J. Murphy, in Advances in Urethane
     Science and Technology, Vol. 2 (K. C. Frisch and S. L.
     Reegen, eds.), Technomic Publishing Company, Westport,
     Connecticut, 1968, p. 221; W. R. Rosemund and M. R.
     Sandner, J. Cellular Plastics 13:182 (1977).

38. L. R. Knodel, U.S. Patent No. 3,865,806 (1975).

39. K. N. Edwards, ed., Urethane Chemistry and Applications,
     Am. Chem. Soc. Symp. Series No. 172, ACS, Washington,
     D.C., 1981.

40. W. T. Diefenback, Linn;s Weekly Stamp News 41 (23):12
     (July 29, 1968).

41. D. K. Ray-Chanduri, J. E. Schoenberg, and T. P.
     Flanagan, U.S. Patent 3,891,584 (1975).

42. L. D. Berger, Jr., and M. T. Ivison, in Water-Soluble
     Polymers (R. L. Davidson and M. Sittig, eds.), Reinhold,
     New York, 1962, p. 169.

43. Anon., Plastic Trends (Sep. 1986), p. 29.

44. A. Anton, U.S. Patent No. 3,575,856 (1971).

45. A. Marzocchi and G. E. Ramel, U.S. Patent No. 3,042,544
     (1962).

46. J. V. Koleske, R. M-J. Roberts, and F. D. DelGiudice,
     U.S. Patent No. 3,725,352 (1973).

47. J. V. Koleske, C. J. Whitworth, and R. D. Lundberg,
     U.S. Patent No. 3,781,381 (1973).

48. J. V. Koleske, R. M-J. Roberts, and F. D. DelGiudice,
     U.S. Patent No. 3,825,620 (1974).

49. POLYOX Water-Soluble Resin Control-Flow Cleaning Solu-
     tions, Union Carbide Corporation, Technical Brochure
     F-42934B.

50. K. E. McConnaughay, U.S. Patent No. 3,110,604 (1963).

50a. F. H. Ancker, U.S. Patent No. 3,377,261 (1968).

51. J. C. Dudduy, U.S. Patent Nos. 3,121,029 (1964) and
     3,181,973 (1965).

52. H. D. Beaumont and A. I. Freidman, U.S. Patent No.
     3,424,605 (1969).

53. C. E. Colwell and R. C. Miller, U.S. Patent No.
     3,020,231 (1961).

54. B. R. Thompson, U.S. Patent No. 3,020,229 (1962).

55. M. B. McGovern, U.S. Patent No. 3,266,888 (1966).

56. J. A. Manley, U.S. Patent No. 3,141,815 (1962).

57. F. E. Bailey, Jr., and J. V. Koleske, Poly(ethylene ox-
     ide), Academic Press, New York, 1976.

58. UCAR™ Rapid Water System, Union Carbide Corporation, Technical Brochure F-43183B.
59. J. F. Eberle, TAPPI 56:10 (1973).
60. U.S. Dept. of Interior, Federal Water Pollution Control Admn., Water Pollution Control Research Series WP-20-22, August 1969.
61. E. S. Mackey, F. Dersch, and R. E. Leary, U.S. Patent No. 3,597,214 (1971).
62. J. F. Kingston, U.S. Patent No. 3,227,942 (1964).
63. J. Schrantz, Ind. Finishing (March 1989), p. 19.
64. C. R. W. Morrison, U.S. Patent No. 3,179,609 (1965).
65. P. L. Layman, Chem. Engr. News (June 1, 1987), p. 19.
66. A. M. Rosenqvist, Studies in Conserv. 4:62 (1959).
67. L. R. Ember, Chem. Engr. News (November 14, 1988), p. 10.
68. J. C. McCawley, J. Can. Conserv. Inst. 2:17 (1977).
69. P. Alagna, Studies in Conserv. 22:158 (1977).
70. A. E. Werner, Studies in Conserv. 6:133 (1961).
71. R. Lefevre, Bull. Inst. Royal Patrimoine Art 3:98 (1960).
72. K. Morris and B. L. Siefert, J. Am. Inst. Conserv. 18: 33 (1978).
73. W. A. Oddy, S. M. Blackshaw, and S. Baker, "The Conservation of Stone," Proceedings of the Intern. Symp., Bologna, Italy, 1976, p. 485.
74. R. M. Orgar, Studies in Conserv. U4:96 (1959).
75. A. J. Stamm, Forest Products J. 6:201 (1956).
76. D. R. Brose, Explorer, Cleveland Museum of Natural History 20:13 (1978).
77. A. J. Klein, Plastics Design Forum (March—April 1989), p. 31.
78. T. W. Sheridan, Plastics Design Forum (March—April 1989), p. 21.
79. R. Lane, J. Immunol. Methods 72:71 (1984).
80. P. H. Silvis, J. W. Walsh, and D. M. Shelow, Am. Laboratory (Feb. 1987), p. 41.
81. Milipore Corp., Polymer Notes 2 (1):1 (July 1987).
82. Anon., Chemical Week (April 19, 1989), p. 52.
83. Anon., Insight (Nov. 6, 1989), p. 58.

Index